Bridging Complexity and Post-Structuralism

Minka Woermann

Bridging Complexity and Post-Structuralism

Insights and Implications

Springer

Minka Woermann
Department of Philosophy
Stellenbosch University
Matieland, South Africa

ISBN 978-3-319-39045-1 ISBN 978-3-319-39047-5 (eBook)
DOI 10.1007/978-3-319-39047-5

Library of Congress Control Number: 2016944360

Printed on acid-free paper

This Springer imprint is published by Springer Nature
The registered company is Springer International Publishing AG Switzerland

For Mila and Stefan

Acknowledgements

A special word of thanks to Sven Woermann, Stefan Vrba, Mari Vrba, Sep Vrba, and Kim Vrba for your love and support; Mila Woermann and Stefan Woermann for the joy and meaning that you bring to my life; my colleagues at Stellenbosch University's Philosophy Department for your collegiality, encouragement, and accommodation; Dr. Rika Preiser for accompanying me on the journey to discovering and articulating the workings of complex systems; and Prof. em. Egidius Berns, Dr. Erik Meganck, and Vasti Calitz for taking the time to read and to comment on an earlier version of this work. I also extend my thanks and gratitude to the late Prof. Paul Cilliers for his intellectual passion and generosity, and for his formative influence on my intellectual development.

Lastly, I thank the *South African Journal of Philosophy* for permission to reprint material on pp. 451–452 of the article titled, 'The ethics of complexity and the complexity of ethics' (co-authored with Paul Cilliers); and on pp. 281–283 of the article titled, 'Moral imperatives for an immanent world: an investigation of the pragmatist, deconstructive, and ante-philosophical positions'.

Contents

Chapter 1
Introduction

> *I… believe that there is at least one philosophical problem in which all thinking men are interested. It is… the problem of understanding the world—including ourselves, and our knowledge, as part of the world.*
>
> - *Karl Popper (1935), The Logic of Scientific Discovery*

Abstract This introductory chapter outlines the aims and scope of the study. In a nutshell, the study presents a sophisticated account of both philosophical complexity and post-structural philosophies of difference. By relating these two fields, an attempt is made to further our understanding of philosophical complexity and to offer an original characterisation of a number of post-structural philosophers as complex thinkers. A brief overview of the post-structural reading of philosophical complexity is also provided. This reading is contrasted with both scientific and critical realist understandings of complexity as these interpretations represent significant theoretical competitors to the post-structural interpretation. The complexity theorists, philosophers, and themes explored in this work are also introduced, and a short synopsis of the various chapters is provided in order to orientate the reader.

1.1 Aims of the Study

Philosophy has the potential to have great explanatory value, both in elucidating that which was previously conceptually veiled, and in providing depth to our understanding of the human condition. To my mind, the hallmark of a successful philosophy is thus related to the degree to which it resonates with our views on, and experiences in, the world. Although this opening expresses sentiments that are perennially affirmed—more often than not in unpromising student essays!—these sentiments nevertheless also explain the underlying drive towards philosophy. In terms of my own schooling, philosophical complexity and post-structuralism (especially the philosophy of Jacques Derrida) have been most valuable to my understanding; and, in a very broad sense, the impetus behind this work has been the desire to articulate and to share this value with others.

M. Woermann, *Bridging Complexity and Post-Structuralism*,
DOI 10.1007/978-3-319-39047-5_1

To this end, the post-structural reading of philosophical complexity that was pioneered by my late colleague and supervisor, Professor Paul Cilliers, is further developed in this study at the hand of the insights of a number of post-structural philosophers and their predecessors. Since the concern is with understanding the world and our place in it, the analyses are framed in terms of economy, (social) ontology, epistemology, and ethics. The theoretical insights gleaned are also illuminated at the hand of the problematic of the foreigner and the related challenge of showing hospitality to foreigners. The reason for focusing on the example of foreigners is because it constitutes one of the central complex human issues of our time in that it leads us to question our conceptions of home and the homeland, of others, and ultimately of ourselves.

In a nutshell, the goals of this study are as follows: I attempt to present a sophisticated account of both philosophical complexity and post-structural philosophies of difference. By relating these two subject fields, I also attempt to both further our understanding of philosophical complexity and offer an original characterisation of a number of post-structural philosophers as complex thinkers. In order to give more content to these aims, this first chapter serves to introduce the various components of this study, and to orientate the reader in terms of the study.

1.2 Philosophical Complexity

Complexity theory is an umbrella term covering many different understandings of, and approaches to, the study of complex systems. In this study, I have used the term *philosophical complexity* to distinguish my understanding of complexity theory from a more scientific understanding. Philosophical complexity has been variously referred to as *soft complexity*, *critical complexity* (Cilliers 2016), and *general complexity* (Morin 2007). Even though each designation signifies a slight shift in emphasis, the hallmark of this approach is that it seeks to explain the world as inherently complex. As noted in Chap. 2, this view thus constitutes a particular view of ontology, rather than a theory of causation (Byrne and Callaghan 2014). Philosophical complexity is contrasted with the scientific understanding of complexity, which is also referred to as *hard complexity* or *restricted complexity* (Morin 2007). Scientific complexity is very much a theory, in that—as articulated in the Santa Fe Institute's mission statement[1]—the focus is on discovering and translating 'the common fundamental principles in complex physical, computational, biological, and social systems' with the aim of formalising and modelling these systems.

Although the goals and underlying assumptions of philosophical complexity and scientific complexity may be very different, these approaches nevertheless share a common history. Tracing this history however proves problematic for the reason that developments do not allow for clear chronological distinctions. Complexity

[1]The Santa Fe Institute is the leading scientific institute for the study of complex systems. The mission statement is available at: http://www.santafe.edu/about/mission-and-vision/.

studies emerged from the interplay of several different disciplines, including physics, mathematics, biology, economy, engineering, and computer science (Chu et al. 2003). Furthermore, its theoretical predecessors include movements such as cybernetics, autopoiesis, General Systems Theory, the theories of artificial intelligence and artificial life, chaos theory, and information theory. In broad terms, the theory of complexity developed during the first half of the twentieth century, when scientists began to study problems of complexity in micro and macro physics (Rasch 1991). In Chap. 2, I offer a partial history of complexity theory in order to situate the field, but also to highlight the ways in which philosophical complexity differs from its predecessors.

A central difference in the development of philosophical complexity and scientific complexity is that philosophical complexity has additionally been influenced by philosophical positions. My approach to philosophical complexity has been largely informed by Cilliers (1998), who, in his seminal work titled *Complexity and Postmodernism: Understanding Complex Systems*, draws on the work of Ferdinand de Saussure, Jean-François Lyotard, and Jacques Derrida in developing an understanding of meaning in complex systems. In this regard, he describes his work as follows:

> The objective of the book is to illuminate the notion of complexity from a postmodern, or perhaps more accurately, *post-structural* perspective. The most obvious conclusion drawn from this perspective is that there is no over-arching theory of complexity that allows us to ignore the contingent aspects of complex systems. If something is really complex, it cannot be adequately described by means of a simple theory. Engaging with complexity entails engaging with specific complex systems (ix).[2]

This description clearly sets philosophical complexity apart from its scientific counterpart, but also from other contesting interpretations of philosophical complexity. Cilliers also sought to distinguish his position through means of his terminology. His term *critical complexity* inaugurates complexity as a primarily reflexive and philosophical (rather than a scientific) account of complexity, but also serves to focus attention on the normativity that any serious engagement with complexity implies. Indeed, as Cilliers' career progressed, he gradually shifted his focus from systematising a complex ontology to exploring the critical consequences that complexity holds. Along with Edgar Morin, Cilliers argues that complexity cannot be resolved through means of a reductive strategy, which is the preferred methodology of those who understand complexity merely as a theory of causation. Instead, critical complexity calls for a thinking that takes seriously the implications that an engagement with complex phenomena holds for our knowledge and social practices.

[2]The distinction between postmodernism and post-structuralism is not hard and fast. Generally, postmodernism is viewed as a response to the ideals of modernism, in which universal abstract principles were sought and contingency avoided; whereas post-structuralism is a response to the structuralist attempt to develop systemic knowledge of language. An overarching aim of these two positions is to debunk as myth the idea of final structure (or a meta-discourse to explain all language forms). Although the positions explored in work are more readily described as post-structural positions, the terms *postmodernism* and *post-structuralism* will be used synonymously.

In practical terms, the critical task amounts to the awareness that a comprehensive, objective description of complex phenomena is impossible. Normative considerations are always at play, both in terms of our modelling choices and the implications that our models hold. Like Derrida's deconstructive philosophy, critical complexity is based on a recursive modality: it is a position that compels us to accept the limits of our knowledge claims, to frame meaning as provisional, and to challenge and to transform our models and practices in a bid to engage responsibly with the complexities of our world. Although I have opted to employ the more general term *philosophical complexity* in this work, the critical and normative implications that Cilliers draws our attention to are central to a post-structural engagement with complexity.

Apart from post-structuralism, a philosophical position that has recently found traction in the complexity discourse is that of critical realism (see Mingers 2006). José López and Garry Potter (2011) describe critical realism as an alternative response to that of postmodernism in regard to 'two significant philosophical developments of the twentieth century: the positivist understanding of natural science and the 'linguistic turn' in the attempt at understanding social phenomena' (6).

As concerns the first development (i.e. the rejection of positivism), the authors note that whilst both postmodernists and critical realists reject positivism (which, crudely-stated, is the belief that science consists of a cumulative and neutral process of discovery), they do so for different reasons. For postmodernists, science is a social construction.[3] Critical realists concede that scientific theories may be constructions, but they argue that there are good rational grounds for believing that one theory gives a better account of reality than another. Although it will be argued in this work (and against López and Potter) that postmodern and post-structural positions are not mute on the question of adjudicating between theories (i.e. these positions are not relativist), it is true that postmodernists and post-structuralists do not endorse the critical realist faith in rationality as the ultimate vehicle via which to navigate between good and bad theories.

As concerns the second development (i.e. the significance of the linguistic turn in understanding social phenomena), López and Potter argue that both postmodernists and critical realists agree that there are significant differences between the social and the natural sciences. These differences are premised on the understanding that 'human society is an object of investigation which possesses features analogous to (or identical with!) language and theory and knowledge are 'language-borne'' (8). However, whereas for postmodernists this conception means that the social sciences cannot be studied using the same methods of the natural sciences (i.e. epistemological causation), critical realists argue that it is *because* of the differences between the natural and the human sciences that a scientific study of the latter is possible. Indeed, 'it is in fact those very peculiarities of the human condition which not only make it amenable to scientific study but actually make social life possible at all' (9).[4]

[3] In this regard, consider Lyotard's (1984) view that any fixing of 'truth' represents a professional self-justification that has the character of a little narrative (*petit récit*).

[4] It should be noted that the view of scientific study forwarded here does not accord with positivism, but is instead based on the acknowledgement that '[s]cience is not pure and can contain an

Although critical realists concede that knowledge is culturally and historically situated, they hold the view that knowledge cannot be reduced to the 'sociological determinants of production' (9). Critical realists therefore believe that we can adjudicate between competing theories 'on the basis of their intrinsic merits as explanations of reality' (9); and, because of this view, critical realism can serve as a philosophical foundation for understanding, and practising, social sciences.

In an article in which critical realism is used as the theoretical framework for investigating the complexities of nursing research, Clark et al. (2008: 67) define the central tenets of critical realism as a 'recognition of reality independent of human perceptions, a generative view of causation in open systems, and a focus on explanations and methodological eclecticism using a postdisciplinary approach'. They argue that these tenets are useful for understanding complex systems and for optimising interventions. This view seems to accord with López and Potter's understanding of the strengths of critical realism, and further seems to support their preference for critical realism over postmodernism, which they argue results in 'a self-defeating relativist scepticism' (López and Potter 2011: 9).

The postmodernist and post-structuralist discourses have been subject to much abuse and misunderstanding. López and Potter's description of postmodernism adequately illustrates the venomous opposition to this philosophical stance. Admittedly a number of postmodernists (the so-called *PoMos*) do subscribe to the 'everything goes' approach. However, sophisticated postmodern or post-structural positions also exist, in which the assumptions underscoring critical realism (especially the strong view of rationality and the faith in [social] scientific progress) are questioned, without foregoing the attempt to explain the nature of reality. Some of these positions, which foreground the contingent aspects of complex systems, and which are best described as co-constructivist rather than constructivist, are explored in the context of this study.

Apart from my schooling, my preference for a post-structural rather than a critical realist understanding of philosophical complexity, lies in the fact that I do not believe that critical realism is an adequate theoretical framework for understanding complexity as a particular view of ontology. Whilst critical realism may facilitate useful interventions in complex systems, I am of the opinion that the critical realist endorsement of a strong view of rationality and epistemological causation, as well as the optimistic belief in (social) scientific progress, mean that this theory is simply not radical enough to affect the revolution that, according to Morin (2007), the intelligence of complexity demands. Viewing complexity as a particular view of ontology, rather than a theory of causation, holds radical implications for our knowledge practices and for our understanding of ourselves and of ethics. These implications need to be dealt with at the fundamental level of theory development, before turning to the question of practical implications.

Therefore, although theories and intellectual positions should have utility value, the utility of this study is not pitched at the level of applying scientific or philosophical insights in order to optimise our interventions in complex systems (i.e. the utility

---

ideologically distorted element in both explanations and the methods used to arrive at them' (López and Potter 2011: 9).

does not lie in the practical application of a particular methodology or set of principles and rules to problems that are deemed complex). Rather, the goal of this study is to attempt to systematise an understanding of the world as inherently complex, and to explore the implications that this view holds for questions of economy, knowledge, identity, and ethics.

Cilliers' *Complexity and Postmodernism* was similarly pitched at this fundamental level, and my hope is that this work can, in some small way, serve as a sequel to his book, and thereby bear testimony to his influence on an emerging generation of complexity thinkers. In this regard, I draw heavily on Cilliers' later works, as well as on the work of Morin, whose profound insights into the nature of complex systems have started to make their way into the English language in recent years. As with *Complexity and Postmodernism*, the philosophical base that undergirds this work is, in part, informed by Saussure and Derrida. However, I also attempt to extend this base by including a number of other thinkers, whose insights have implications for our understanding of complexity. As such, this work incorporates both recent developments in the field of complexity, and builds on the post-structural understanding of complexity that was first introduced in Cilliers' work.

1.3 Post-structuralism

The field of post-structuralism is marked by a great diversity, which makes it difficult to defend the choice of philosophers selected for this work. Compounding the issue is the fact that very few (if any) of the thinkers associated with post-structuralism sit comfortably within the field. Indeed, whilst some straddle fields (such as Derrida, who is associated with both the fields of post-structuralism and phenomenology), others emphatically resist categorisation within a field (in this regard, consider Michel Foucault (2002: 19), who famously stated in the introduction to the *Archaeology of Knowledge*, 'Do not ask me who I am, and do not ask me to remain the same').

Nevertheless, I do believe that there is a thread that ties the philosophers discussed in this work together. In every thinker discussed, the significance of the non-closure of meaning and the impossibility of tracing thoughts or being back to an alleged origin is recognised, the contingent nature of knowledge and identity is stressed, and ethics is construed in terms of praxis or engagement with the complexity generated by the non-closure of meaning.

The primary thinkers guiding this study are Jacques Derrida and Jean-Luc Nancy. As previously stated, Cilliers was greatly influenced by Derrida in developing his insights into the nature of complex systems. In this work, I also tap into the depth and richness of Derrida's oeuvre in order to further tease out the value that his insights hold for our understanding of philosophical complexity. Nancy's work is a new addition to the post-structural understanding of philosophical complexity. Chronologically speaking, Nancy succeeds Derrida, and although he exhibits '"deconstructive" commitments wherever he addresses the limitations of questioning, the demands of theory or the requirements of a given discourse' (Hutchens 2005: 32), it would be incorrect to think of him as a Derridean.

Both Derrida and Nancy support the view that meaning cannot be exhausted. However (and going much too fast), whereas Nancy demonstrates the non-closure of meaning by focusing on the manner in which we are continually exposed to existence in community with others (he speaks of being as *being-in-common*), Derrida illustrates the non-closure of meaning by demonstrating how reified structure or signified content is always subject to the workings of deconstruction. Both these levels of analysis yield interesting—although different—implications for our understanding of complexity, which are explored over the course of the work.[5]

Apart from these primary thinkers, selected views of a number of other philosophers are also drawn on in order to give content to the specific themes of economy (Chap. 3), knowledge (Chap. 4), and identity (Chap. 5). Specifically, Georges Bataille's understanding of a restricted and a general economy are discussed in Chap. 3 in order to shed light on the economic and aneconomic dimensions of meaning; Foucault's archaeological and genealogical methodologies are explored in Chap. 4, with the goals of demonstrating both how knowledge is contingent on the rules of formation guiding a given epoch, and why knowledge is always the product of a complex genealogy of meaning; and, Emmanuel Levinas's view of the relation between the self and the Other are introduced in Chap. 5, since this view gave rise to the recognition and significance of otherness that serves as the grounding for a highly influential philosophy of difference, which arguably came to fruition in the work of Derrida.

Where necessary for elucidating their arguments, the philosophical influences on these post-structural thinkers are also expounded upon. In this regard, the work of Immanuel Kant, Georg Wilhelm Friedrich Hegel, Friedrich Nietzsche, Ferdinand de Saussure, and especially Martin Heidegger are also referenced in this study.

1.4 Methodology

The transdisciplinary nature of this study, as well as the number of philosophical positions referred to, necessitate a word on methodology. This study is primarily exploratory in nature, which means that the goal is not to argue for a specific position, but rather to explore a number of positions, with the aim of integrating the insights derived into our understanding of philosophical complexity. Whereas the aim of Chap. 2 is to introduce the reader to the field of philosophical complexity, the aim of Chaps. 3–6 is to explore philosophical positions that are sensitive to complexity. In Chap. 7, the

[5] As the primary thinkers guiding this work, Nancy's and Derrida's insights are discussed in relation to each of the themes addressed. The only exception concerns Chap. 4, which deals with the question of knowledge. Nancy's work is not dealt with in this chapter, and the reason for this omission is because signified content or the symbolic-ideological order, which is the proper domain of knowledge, does not constitute his philosophical focus.

insights derived from the foregoing chapters are summarised in the language of complexity, and the manner in which the analyses have extended our understanding of philosophical complexity is demonstrated. As such, Chaps. 2 and 7 treat philosophy in terms of complexity, and Chaps. 3–6 treat complexity in terms of philosophy.

As regards the philosophical component of the study (i.e. Chaps. 3–6), and in keeping with the goals of the study, the various philosophical positions are elucidated by bringing these positions into debate with one another. In so doing, the development of thought on a given topic is demonstrated, as are the differences between the philosophical positions explored. Again, it is important to reiterate that, although critical comparisons are undertaken, the point is not to try to defend one position over another, but to tease out the implications that the various analyses hold for our understanding of philosophical complexity.

1.5 The Ethics of Living and the Question of the Foreigner

A central insight that is derived from exploring philosophical complexity and philosophical positions that are sensitive to complexity is that ethics is unavoidable. However, it is important to note that within the context of this study ethics does not denote a normative system dictating right action, but concerns the unavoidable normative dimension of being together with others in a world defined by complexity. In other words it is *because* of complexity, or the non-closure of meaning, that our decisions and actions cannot be objectively described. Instead, we must engage in contingency, alterity, and the over-determinations that characterise our contexts (all of which involve judgement and sense-making that surpass calculation and pure rational argumentation). *The ethics of living*[6] thus constitutes a praxis through which we are continuously being formed, and through which we co-constitute one another and the world and form a sense of our embeddedness in contexts.

Despite the fact that this work is primarily an exercise in theory building, the example of the foreigner, stranger, or intruder is drawn on in order to elucidate the implications that the theoretical analyses hold, specifically as concerns the ethics of living. Although the issue of showing hospitality to the other (that is, to unknown foreigners and to unknowable strangers) is analysed at length, the question of the foreigner also introduces the broader problematic of otherness and of understanding the self as other. This links with the post-structural thematic of the denial of unity as an ontological or epistemological imperative. In this regard, Derrida (2000: 3) writes that the foreigner question is 'the question of being-in-question'; the for-

[6]Note that in this work, the phrase *the ethics of living* denotes the task of living a good life. The phrase in no way refers to a metaphysical account of what living or life as such entails.

eigner is the one 'who… puts me in question'. E.E. Cummings (1965) captures this insight beautifully in the following poem:

a total stranger one black day
knocked living the hell out of me--
who found forgiveness hard because
my(as it happened)self he was
-but now that fiend and i are such
immortal friends the other's each

As concerns the first line of inquiry, it is clear that the social, ethical, and political implications raised by the question of the foreigner, as well as the related ethical imperative of showing hospitality towards foreigners, are currently at the forefront of the global agenda. The question of the foreigner looms large in my country. Indeed, the xenophobic attacks on African foreigners in South Africa in May 2008 are all too fresh in the nation's psyche, and many foreigners living in South Africa today continue to feel fearful. Internationally, we see a surge in nationalism and fundamentalism, both of which promote extreme intolerance to any form of difference and alterity. The 2015 attack on the satirical magazine *Charlie Hebdo's* office in Paris, and the debates on freedom of expression and religious tolerance that ensued, illustrate both the evil resulting from extreme intolerance and the difficulties involved in defining the levels and limits of tolerance. Apart from tolerance, foreigners also raise issues related to hospitality. The Syrian refugee crisis that began in 2011, and that had resulted in the displacement of almost four million people by 2015, poses difficult questions for countries that are under pressure to accommodate these asylum seekers. The spate of terror attacks that have rocked the world of late (and that seem unlikely to abate in future) compound the difficulties in defining these issues, as people feel that their countries and cultures are increasingly threatened by an outside force. There are no easy solutions to these complex problems, but a sustained analysis of the issues can hopefully help us to better understand some of the dimensions that define these issues.

The second line of inquiry, namely the identity challenges introduced by a focus on foreigners and otherness, holds a personal interest for me. This interest was piqued by a confrontation with a stranger a number of years ago (which I relate in Chap. 5), to which I unfortunately did not respond with good grace. This confrontation raised several questions for me, including what the appropriate response to the stranger, who is perceived as threatening, should be; and how one's engagement with otherness serves to influence—and to challenge—one's understanding of self.

The question of the foreigner has also been dealt with at length by both Derrida and Nancy. Their insights are drawn upon to give philosophical depth to the illustration, which is treated in three parts. In Chap. 4 the difficulties involved in defining (in complexity terms, modelling) the foreigner, who by definition resists appropriation into our conceptual schemas, is explored at the hand of Derrida's work. The implications and difficulties that this holds for acting hospitably towards foreigners is also addressed. In Chap. 5, the Levinasian-Derridean view of the manner in which one's identity is always intertwined with the identity of the foreigner is investigated, as is the Nancean view that the foreigner (what Nancy calls the intruder) resides at the very heart of identity. Indeed, Nancy's ontology poses a view of a divided self; a self who is always already both self and other, singular and plural. In Chap. 6, I attempt to give some content to what an ethics of living may entail by drawing on the writer and scholar, Jonny Steinberg's portrayal of actively pursuing foreignness as a means of remaining open to alterity, whilst living a well-defined and ordered life.

By drawing on the example of the foreigner, the hope is that the insights and challenges related to a complex view of knowledge creation, identity formation, and ultimately ethical living become tangible to the reader; and that, in this process, the value of philosophical complexity and complex philosophy may become clear.

1.6 Chapter Synopses

The above introductory discussion was intended as a general overview of the approach taken to the study, and the content covered in the study. In this final section, a more comprehensive chapter synopsis is provided.

The term *complexity* is often loosely appropriated by both academics and practitioners to describe things that lack simple explanations. However, as previously argued, little conceptual clarity exists regarding the meaning of the term due, in part, to the heterogeneous history of complexity theory. In Chap. 2, a partial account of this history is offered. To this end, an overview of first-order cybernetics and information theory, second-order cybernetics and autopoiesis, third-order cybernetics and artificial life, and General Systems Theory is presented. The influence of these theoretical predecessors on philosophical complexity is highlighted, as are the central points of divergence. In so doing, a description of the features of philosophical complexity emerges. Philosophical complexity is also compared and contrasted with the scientific school of complexity theory. The goal is to illustrate why philosophical complexity—which, as previously stated, can best be described as presenting a particular view of ontology—necessitates a different methodological approach to that of scientific complexity.

The ontological view of complexity is the view that the world is inherently complex, because there is no central organising principle and the system is open. In Chap. 3, this notion of an inherently complex and open system is reinscribed and further explored in terms of a philosophical understanding of economy. The notion of economy, which denotes any constrained set of relations, was introduced into the philosophical literature by Bataille. In his work, he distinguishes between a utilitarian or restricted economy and an excessive or general economy. In complexity terms, Bataille's central insight is the following: complexity is generated by a constrained set of relations, which gives rise to a system (the restricted economy of codified knowledge), but also to excess. The excess forms part of the system's environment (the general economy).

The exact manner in which the relation between the restricted and the general economies, or the system and its environment, should be conceptualised forms the grounds for different conceptions of economy. In this chapter, five views on economy are forwarded, namely Hegel's totalising economy, Nietzsche's and Bataille's dual economy, Derrida's *aporetic* economy, and Nancy's immanent economy. The implications that these positions hold for understanding both systemic openness and relationality are explored at the hand of critical analyses in order to provide a philosophical account of the ontological view of complexity.

In Chap. 4 the epistemology of complexity is explored. This problematic can be summarised as follows: the only complete description of a complex system is the system itself. However, since we cannot understand complexity in all its complexity, we are forced to model complex systems in order to render these systems meaningful. Models therefore reduce complexity, and are both an outcome of our mental constructions and empirical considerations. Complexity thinking necessitates an engagement with the nature, limits, and consequences of knowledge production, which raises both technical and ethical considerations.

This complexity-based view of knowledge is further explored in this chapter at the hand of philosophical positions that are sensitive to the normative dimension of knowledge creation. Specifically, Foucault's work on archaeology and genealogy is introduced in order to demonstrate how our knowledge practices are informed by largely unconscious epistemic frames, are construed on the basis of unreliable premises, and develop according to complex historical processes. The implications that Derrida's deconstructive philosophy hold for drawing attention to, and dealing productively with, the limited status of our models and the impossibility of the closure of meaning are also explored. Both these philosophical positions reinforce the view that the epistemology of complexity cannot be construed as a value-free programme. The very real consequences that this position holds are explored at the hand of the question of the foreigner, who we seek to understand, but who ultimately resists assimilation into our conceptual schemas.

The epistemology of complexity holds important implications not only for the meaning of our knowledge practices, but also for our understanding of identity. In Chap. 5, a complex view of identity, as well as the ethical-political implications that such a view gives rise to, are explored. Philosophical complexity and post-structuralism both offer a challenge to the Cartesian humanist subject and the predicates of subjecthood that this view presupposes (including a strong view of agency, intentionality, rationality, and causality). In substantive terms, this challenge results in a so-called *liquidated subject*, wherein the notion of the subject no longer corresponds with any fixed or signified content, but is instead characterised as a decentred and complex construction.

In this chapter, the deconstruction of the humanist subject, and the traditional predicates of subjecthood, is undertaken at the hand of Levinas's understanding of the Other; Derrida's work on the subject, animals, and eating; and, Nancy's reading of Heidegger's *Dasein* from the perspective of *Mitsein*. The insights that these analyses yield are critically compared to Cilliers' complex view of identity, in which the self becomes over time in a network of relations with others. In so doing, a tentative portrait of the liquidated subject emerges, and attention is drawn to the urgent need to revise our traditional understanding of ethics and responsibility. This latter point is explored in terms of the interplay between foreignness and identity.

Chapter 6 presents both a summary of the ethical insights gleaned from the analysis, and more detail on the ethical position espoused in this work. Specifically, it is argued that both the philosophical and complexity positions that are drawn on necessitate an ethics of living, understood in terms of ethical praxis. The exact nature and significance of ethical praxis, however, varies according to the position studied. Two views are presented, namely Derrida's experiential praxis and Nancy's praxis defined as an engagement in, and with, originary ethics. Whereas Derrida's view of praxis gives rise to an ethics of alterity, Nancy's praxis translates as the duty to make sense or the duty to respect existence. Their insights are translated into the practical implications that the analysis holds for understanding, and practising, the ethics of living.

In the concluding chapter, the notion of philosophical complexity is revisited and further explored in terms of the prominent philosophical insights gleaned over the course of the study. It is argued that systemic openness is the defining feature that distinguishes complex systems from non-complex systems. As such, the concept of *opening* constitutes the theoretical frame through which the post-structural insights are interpreted and translated in this chapter. Opening gives rise to three paradoxes that define complex systems, namely that these systems are both ordered and disordered (and are constantly in a state of decay and renewal); complex systems are autonomous from, as well as dependent on, the environment; and, the identity of complex systems is both exclusive and inclusive. Each of these paradoxes is explored at the phenomenological level.

1.7 Conclusion

Bridging complexity and post-structuralism admittedly makes for demanding reading because it requires of the reader to jump between language games, and because the analyses presented refuse to deliver neat and complete conclusions. Yet I believe that this task is important, since—to cite my favourite line by Derrida (1988: 119)—'If things were simple, word would've gotten around.' Through this work, I aim to make a convincing case for my belief that a robust understanding of today's world requires that we grapple with complexity and the implications that it holds; and, moreover, that this task is not only a theoretical exercise, but an ethical imperative. In this regard, Zygmunt Bauman's (1993: 15) defence of postmodernism applies equally to philosophical complexity: the insights yielded by both positions are unlikely to make life easier, but they can provide substance to the dream of making life 'a bit more *moral*.'

References

Bauman, Z. 1993. *Postmodern ethics*. Oxford: Blackwell.

Byrne, D., and G. Callaghan. 2014. *Complexity theory and social sciences: The state of the art*. Oxon: Routledge.

Chu, D., R. Strand, and R. Fjelland. 2003. Theories of complexity: Common denominators of complex systems. *Wiley Periodicals* 8(3): 19–30.

Cilliers, P. 1998. *Complexity and postmodernism: Understanding complex systems*. London: Routledge.

———. 2016. *Critical complexity: Collected essays*, ed. R. Preiser. Berlin: De Gruyter/Ontos.

Cummings, E.E. 1965. *A selection of poems*. New York: Harcourt.

Clark, A., S. Lissel, and C. Davis. 2008. Complex critical realism: Tenets and applications in nursing research. *Advances in Nursing Science* 31(4): 67–79.

Derrida, J. 1988. Afterword. In *Limited Inc,* ed. G. Graff, trans. S. Weber, 111–160. Evanston: Northern Western University Press.

———. 2000. *Of hospitality*. Trans. R. Bowlby. Stanford: Stanford University Press.

Foucault, M. 2002. *The archaeology of knowledge*. Trans. A.M. Sheridan Smith. London: Routledge Classics.

Hutchens, B.C. 2005. *Jean-Luc Nancy and the future of philosophy*. Chesham: Acumen.

López, J., and E. Potter. 2011. After postmodernism: The millennium. In *After postmodernism: An introduction to critical realism*, ed. J. López and E. Potter, 1–16. London: The Athlone Press.

Lyotard, J.-F. 1984. *The postmodern condition: A report on knowledge*. Trans. G. Bennington and B. Massumi. Minneapolis: University of Minneapolis Press.

Mingers, J. 2006. *Realising systems thinking: Knowledge and action in management science*. New York: Springer.

Morin, E. 2007. Restricted complexity, general complexity. Trans. C. Gershenson. In *Worldviews, science and us: Philosophy and complexity*, ed. C. Gershenson, D. Aerts, and B. Edmonds, 5–29. Singapore: World Scientific.

Popper, K. 1935/2005. *The logic of scientific discovery*. Taylor and Francis e-library. http://strange-beautiful.com/other-texts/popper-logic-scientific-discovery.pdf. Accessed 13 March 2015.

Rasch, W. 1991. Theories of complexity, complexities of theory: Habermas, Luhmann, and the study of social systems. *German Studies Review* 14(1): 65–83.

Chapter 2
Towards an Understanding of Philosophical Complexity

The ideal art, the noblest of art: working with the complexities of life, refusing to simplify, to "overcome" doubt.

— Joyce Carol Oates, The Journal of Joyce Carol Oates: 1973-1982

Abstract
The term complexity is often loosely appropriated by both academics and practitioners to describe things that lack simple explanations. However, little conceptual clarity exists regarding the meaning of the term. This is partly due to the diverse history of complexity theory, which evolved from the interplay of several disciplines.

In this chapter, a partial account of this history is offered. To this end, an overview of first-order cybernetics and information theory, second-order cybernetics and autopoiesis, third-order cybernetics and artificial life, and General Systems Theory is presented. The influence of these theoretical predecessors on philosophical complexity is highlighted, as are the central points of divergence. In so doing, a description of the features of philosophical complexity emerges.

Apart from its heterogeneous history, complexity theory currently refers to a number of theoretical enterprises, based on different assumptions, methodologies, and aims. In this chapter, philosophical complexity is also compared and contrasted with the scientific school of complexity theory. The goal is to illustrate why philosophical complexity—which can best be described as presenting a particular view of ontology, rather than a theory of causation—necessitates a different methodological approach to that of scientific complexity.

2.1 Introduction

Complexity is undoubtedly one of the buzzwords of our time. The French complexity theorist, Edgar Morin (2008: 19), argues that in popular parlance complexity has traditionally been understood as a term which 'always carried with it a warning to our understanding, a cautioning against clarification, simplification, and overly

M. Woermann, *Bridging Complexity and Post-Structuralism*,
DOI 10.1007/978-3-319-39047-5_2

rapid reduction.' This meaning is also reflected in the above opening quote by Joyce Carol Oates, whose fiction is characterised by an unflinching grappling with human complexity. Her goal is not to overcome the complexity, but to help the reader to reconcile herself with a portrait of humanity that bursts forth from the neat frames that we as a society employ in order to define, to categorise, and to judge our fellow man. If the goal of art is to interpret the human condition, then Oates's work is indeed a poignant illustration of how complexity defines the very heart of who we are.

Apart from this popular understanding of complexity, the term has also been employed in academia in an attempt to develop a systematic body of knowledge pertaining to complex phenomena. However, and as noted in Chap. 1, the disciplines in which complex phenomena are studied are very diverse, as are the founding assumptions on which the study of complexity is based. In other words, as a theoretical enterprise, there is little uniformity regarding the definition and characteristics of complexity theory. In this work, I attempt to develop a philosophical understanding of complexity, which is presented as an ontologically-founded framework for understanding the world as constituted primarily of complex systems (Byrne and Callaghan 2014). As a philosopher, I am however also interested in the phenomenological implications that such a framework holds for our experiences of, and in, the world. I therefore also attempt to develop our knowledge of the human condition understood from the vantage point of a complex understanding of the world. In other words, my goal is to marry a popular and a more theoretical understanding of complexity, in which the experience of complexity is reconciled with an understanding of the world as inherently complex. In so doing, I hope to illustrate why the ontological understanding of complexity, as well as the phenomenological implications that this ontology implies, necessarily commit us to an ethical position.

The goal of this particular chapter however, is to clarify the understanding of philosophical complexity that will be employed in this work. This is achieved by tracing the development of philosophical complexity, and by presenting an overview of the features and methodological commitments of philosophical complexity. Two significant movements that have impacted on our understanding of complexity are explored, namely cybernetics and General Systems Theory (GST).

In this regard, I follow Morin (1992a), who notes in the foreword to the first volume of his six-part magnum opus, titled *Method*, that he initially treated the problem of organisation (which for him is the starting point for conceiving of complexity) at the hand of GST and cybernetics. In so doing, he discovered the following:

> En route, these ideas, of solutions, became starting points, then finally scaffolding, necessary certainly, but to be dismantled after having raised us to the concept of organization. Starting at a certain stage then, these liberating ideas locked me in. I was able to develop their message only by metamorphosing them… It seems to me [thus] that systemic and cybernetic (including information) ideas are integrated here, that is to say preserved in their sap and their truth, but at the same time provincialized, criticized, transformed, complexified (23).

In tracing the history of complexity theory, my learning was informed by Morin's experience. Although I neither pretend to understand nor to explain this history in the same depth and with the same proficiency as Morin does, it is nevertheless a worthwhile task to outline the history of complexity. This is because this history sheds light on the principles of complexity, by highlighting how these principles resonate with (or diverge from) important insights gleaned from both cybernetics and GST. Apart from juxtaposing complexity insights with these movements, philosophical complexity is also explored and contrasted with the key assumptions on which complexity science (introduced in the previous chapter) is based. This is done in order to present a fuller account of the methodological underpinnings of philosophical complexity.

2.2 Cybernetics

The cybernetics paradigm—which, as noted by Hans von Foerster (1990), is variously described in terms of 'goal-orientated behaviour' (Norbert Wiener), 'feedback' (Margaret Mead), 'control, recursiveness and information' (Gregory Bateson), 'the science of effective organisation' (Stafford Beer), and 'the science of defensible metaphors' (Gordon Pask)—has played a huge role in clarifying the manner in which complex systems function. In order to provide a chronology of ideas, the following discussion treats cybernetics in terms of Katherine Hayles's (1994, 1999) categorisation of first, second, and third-order cybernetics.

2.2.1 First-Order Cybernetics and Information Theory

The field of cybernetics has its origins in the techno-scientific project that was implemented during World War II by the American government (Lafontaine 2007). Cybernetics also served as the precursor to robotics and computer information ethics. Wiener, a prominent mathematician, is generally considered to be the founding father of cybernetics. He also originally coined the term *cybernetics*, which is derived from the Greek word *kybernētēs* (which means steersman, governor, pilot, or rudder). However, many influential thinkers were involved in the establishment of cybernetics as a field, including the mathematicians, John von Neumann and Warren Weaver; the engineer, Claude Shannon; the physicist, Heinz von Foerster; and the cultural anthropologists Margaret Mead and (her third husband) Gregory Bateson.

Céline Lafontaine argues that—despite the interdisciplinary nature of the field—cybernetics is based on three major principles, namely entropy, information, and feedback. These principles lay the foundation for a new science. Lafontaine further notes that, although important advances in cybernetics were made during the wartime years, it was only at the end of the war that some of the above-mentioned acclaimed thinkers met to discuss questions of control and feedback. The forum for the discussion was the *Macy Conferences on Cybernetics*, which were held between 1946 and 1953 (first-order cybernetics dates from 1945 to 1960) (Hayles 1994: 441).

The first *Macy Conference*, which was led by von Neumann and Wiener, marked the dominance of the machine-concept and the triumph of 'information over materiality' in that 'the important entity in the man-machine equation was [identified as] information, not energy' (Hayles 1999: 51). Apart from arguing for the construction of information as a theoretical entity, Hayles explains that members of the *Macy Conferences* also sought to construct both (human) neural structures as flows of information and artifacts in a manner that translated information flows into observable operations.

Morin (2008) explains that information emerged with Ralph Hartley, and especially Shannon and Weaver, as communicational on the one hand, and statistical on the other. The view of information as communicational was pioneered by Shannon, and was integrated into a theory of communication dealing with the transmission of messages. Morin (1992a: 235) notes that in this theory, communication is conceived of in organisational terms because it is 'an organizational link effected by the transmission and exchange of messages'. This, he contends, marks '[t]he first originality of cybernetics'. The view of information as statistical deals with the probability of an elementary unit of information appearing. The concept of *entropy*—which refers to the level of disorder in a system—gained significance at the beginning of the twentieth century, with the formulation of the second law of thermodynamics, which states that the universe tends towards general entropy or maximal disorder (Morin 2008).

The link between this Shannonian equation of information and entropy lies in the discovery that entropy is inversely proportional to information. Or, otherwise stated, 'there is equivalence between information and negative entropy or negenentropy' (14). This is illustrated in Fig. 2.1., which shows the process that produces a message. Shannon (1948) writes that such a process consists of (1) an information source, which produces the message that is to be communicated to the receiver; (2) a transmitter, which operates on the message in order that it may be transmitted over a channel; (3) a channel, which is the medium for transmission; (4) a receiver, who reconstructs the message from the signal; and, lastly, (5) a destination, which signifies the person or thing for whom/which the message was intended. The success with which the signal is received, is contingent on noise perturbations (i.e. entropy) acting on the communication system.

Not only is this theory relevant to the communication of messages on a computer, but information theory could also be extrapolated to the biological realm, whereby, for example, genetic mutation is likened to noise or entropy that disrupts the reproduction of a message (information), thereby provoking an 'error' (with respect to the original message) in the constitution of a new message. The novel element is therefore that information theory 'could, on the one hand, be integrated into the notion of biological organization… [and], on the other hand, somewhat surprisingly link thermodynamics, or physics, to biology' (Morin 2008: 13).

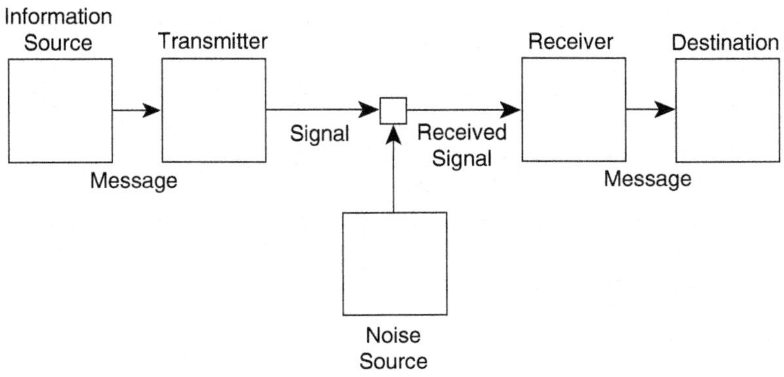

Fig 2.1 'Schematic diagram of a general communication system'(*source*: Shannon 1948: 381)

This new science also marks an epistemological revolution in that the intrinsic study of beings and objects is rejected in favour of an approach that focuses on 'interactions between objects, regardless of their nature (physical, biological, artificial or human)' (Lafontaine 2007: 29). In this sense, cybernetics poses a challenge to the distinction between human and machine, since the behaviour of both thermostats and people, for example, can be explained 'through theories of feedback, hierarchical structure, and control' (Hayles 1999: 84; see Haraway 1985). In this regard, Morin (1992a: 248) argues that the central strength of this paradigm lies in the fact that:

> Cybernetics is the first science… to have founded its method, effected it operational success, and made itself recognized by the other sciences by envisioning a physical system, the machine, not in function of its constitutive elements, but in function of its organizational characters.

Morin (235) further notes that '[t]he second originality of cybernetics is to link informational communication and command'. He thus defines cybernetics as 'the theory of the command… of systems whose organization includes communication.' Articulating the manner in which to control and pilot communication systems has been instrumental in computer programming and in developing a range of computational operations and performances. However, Morin argues that the focus on the machine, coupled with the emphasis on command, marks the central weakness of theory, which is that ultimately communication is subordinated to command, and moreover, to the command of the machine. The consequences arising from this view are the following:

> There is neither essence (which is an advantage) nor existence (which is a deficiency) in the cybernetic grasp of the living being, which becomes very serious as soon as cyberneticism claims to interpret life, man, society… [T]he subordination of communication to command not only prevents cybernetics from conceiving the relation communication/command in its generic complexity, but forces it to conceive biological organization and social organization only as enslavement (251).

As such, Morin identifies this movement as essentially reductionist in that the vast anthropo-social sphere is reduced to 'the vision of an engineer on machines', with the consequence that all of life is swallowed up in a 'model of self-styled rationality: the automated, functionalized machine, purged of all disorders… end-purposed for industrial production' (253). In a bid to avoid this dystopian vision, Morin argues that we should reverse the sovereignty of control in favour of communication. His hope is therefore that communication commands, since it is communication—not control—that should define the heart of our anthropo-social condition. If communication commands, then communication can take on its fullest expression, which, according to Morin, is individual, social, political, and ethical. In this context, humans—not machines—remain in control.

The reintroduction of human agency was, to a large extent, facilitated by the second movement of cybernetics, in which the idea of scientific objectivity (or one master formula to solve all problems through means of simplification, reduction, and manipulation) is debunked in favour of a reflexive view of knowledge creation. The question of reflexivity already arose during the *Macy Conferences*. Whilst some of the physicists steered the debate to more comfortable grounds by continuing to endorse the old paradigm, a group of thinkers (led by von Foerster) raised the problem of reflexivity (Hayles 1999). The central issue that preoccupied this group was 'how to redefine homeostatic systems so that the observer can be taken into account' (10).

The foundational stage of cybernetics is therefore characterised by the emergence of two constellations, which were in competition with each other, namely: a conservative constellation that privileged 'constancy over change, predictability over complexity, [and] equilibrium over evolution' (Hayles 1994: 446); and a second constellation that privileged 'change over constancy, evolution over equilibrium, [and] complexity over predictability' (446). Whereas the conservative constellation focused on the concept of *homeostasis*, defined as 'the ability of an organism to maintain itself in a stable state' (446), the central concept that interested researchers from the second constellation was reflexivity. Hayles (446) defines reflexivity as 'turning a system's rules back on itself so as to cause it to engage in more complex behavior.'

2.2.2 Second-Order Cybernetics and Autopoiesis

During the 1960s second-order cybernetics developed in an attempt to account for the observers of systems, who are themselves also systems. The initiative was driven by von Foerster, who experimented with various ways in which to formulate reflexivity.[1] Von Foerster (in Von Foerster and Poerksen 2002: 110) describes the difference between first-order and second-order cybernetics as follows:

[1] In this regard, see von Foerster's (1984) influential book, titled *Observing Systems*.

> First order cybernetics separates the subject from the object and refers to an assumed world "out there". Second order cybernetics or cybernetics of cybernetics is itself circular. You learn to understand yourself as a part of the world that you wish to observe. The entire observational situation ends up in another area in which you suddenly have to take responsibility for your observations.

In a plenary presentation, titled, 'Ethics and second-order cybernetics', von Foerster (1990) again defines circularity as the central theme of cybernetics. He, moreover, argues that this circularity pertains not only to the subject matter of cybernetics, but also to the manner in which this subject matter is investigated, and the manner in which the researcher begins to view him or herself ('as being included in a larger circularity'). In short, circularity implies that 'the cybernetician, by entering his own domain, has to account for his own activity; cybernetics becomes cybernetics of cybernetics, or second-order cybernetics'.

Von Foerster argues that this circularity in scientific investigation is self-evident. This is because if 'the properties of the observer, namely to observe and to describe are eliminated, there is nothing left: no observation, no description.' And yet, this self-evident truth has also met with much resistance, for the reason that circularity breeds paradoxes, and the threat of paradoxes stealing their way into theory is, according to von Foerster, 'like having the cloven-hoofed foot of the Devil stuck in the door of orthodoxy [and objectivity]'. Von Foerster describes the paradoxes of circularity as follows:

> *In the general case of circular closure, A implies B, B implies C, and—O! Horror!—C implies A!*
> Or the reflexive case:
> *A implies B, and—O! Shock!—B implies A!*
> And now Devil's cloven-hoofed foot in its purest form, in the form of self-reference:
> *A implies A.*
> —Outrage!

Although von Foerster did not fear the loss of scientific orthodoxy and objectivity that circularity implies, he did grapple with the problem of how to speak about reflexivity without sliding into a solipsistic position. Of this problem, Hayles (1999: 133–134) writes: 'The message from the *Macy Conferences* was clear: if reflexivity was to be credible, it had to be insulated against subjectivity and presented in a context which had at least the potential for rigorous (preferably mathematical) formulation.'

According to Hayles, a breakthrough occurred in 1969 when the Chilean biologist, Humberto Maturana, unveiled his ideas of treating cognition as a biological phenomenon at a conference to which von Foerster had invited him. This idea was presented in the seminal paper, titled 'What the frog's eye tells the frog's brain', in which Maturana and his co-authors demonstrate 'that the [frog's] eye speaks to the brain in a language already highly organized and interpreted instead of transmitting some more or less accurate copy of the distribution of light upon the receptors' (Lettvin et al. 1959: 1950). In other words, instead of registering reality, the frog's perceptual system constructs reality (Hayles 1999). According to Hayles (1994: 461), this article—which discredits the idea of a neutral vantage point from which to observe reality—served to blow 'a frog-sized hole in objectivist epistemology.' Maturana makes similar observations in his work on colour vision, in which he demonstrates that there is no direct correlation between an animal's colour perception and the world. Rather, a correlation exists between the activity in an animal's retina and its experience of colour (Hayles 1999; see Maturana et al. 1968). Both his work on the frog's perceptual system and colour vision, led Maturana to conclude that there can be no unmediated understanding of reality. Rather, as Hayles (1999: 136) states, '[reality] comes into existence for us, and all living creatures, *only through the interactive processes determined solely by the organism's own organization.*'

Hayles argues that second-order cybernetics reached its mature stage with the publication of Maturana and his co-author, Francisco Varela's (another Chilean biologist's), influential book titled *Autopoiesis and Cognition: The Realization of the Living* (1980). Their central premises are that all systems are autonomous, self-referential, and operationally or organisationally-closed. In other words, 'living systems operate within the boundaries of an organization that closes in on itself and leaves the world on the outside' (Hayles 1999: 136). This means that systems are only open to the environment from an observer's perspective. Therefore, '[e]verything said is said by an observer' (Maturana and Varela 1980: xxii). This however does not mean that organisationally-closed systems are isolated systems (i.e. systems that are completely isolated from their environments). The point is rather that the environment is drawn into the system in order to facilitate its own production and maintenance (Morgan 2006: 244). Roberto Poli (2009: 8) explains the consequence of this view as follows: 'the system's connection with the environment is no longer a kind of immediate and direct relation between the system and its environment but becomes a reflexive relation, mediated by the self-referential loops that constitute the system itself.' Self-reference is thus the key to understanding organisationally-closed systems. To illustrate this point, consider Maturana and Varela's (1980: xv) treatment of the nervous system as an activity 'determined by the nervous system itself and not by the external world'. The external world is only accorded 'a triggering role' in releasing 'the internally-determined activity of the nervous system.'

Maturana and Varela further maintain that, because systems are organisationally-closed, their ability to self-create or self-renew is due to the system's capacity for self-production through feedback loops—a process which they call *autopoiesis*. In *Autopoiesis and Cognition*, Maturana explains the origins of the word autopoiesis: he and Varela initially used the term *circular organisation* to refer to the system's capacity for self-production through feedback loops, but found this term to be inadequate, since it did not convey the idea of autonomy that they identified as the central characteristic of the organisation of the living. The term *poiesis* followed after a conversation with a friend regarding 'Don Quixote's dilemma of whether to follow the path of arms (*praxis*, action) or the path of letters (*poiesis*, creation and production)' (xvii). It was after this conversation that Maturana claims to have understood 'for the first time the power of the word 'poiesis' and thus 'invented the word that [they] needed: *autopoiesis*' (xvii). A few pages on, Maturana and Varela (9) provide the following description of autopoiesis:

> It is the circularity of its organization that makes a living system a unit of interactions, and it is this circularity that it must maintain in order to remain a living system and to retain its identity through different interactions.

What differentiates the reflexivity in autopoietic systems from the understanding accorded to it in the *Macy Conferences* is that reflexivity is no longer associated with 'psychological complexity', but is rather viewed as being 'constituted through the interplay between a system and its components… [which] mutually define each other in the bootstrap operation characteristic of reflexive self-constitution' (Hayles 1994: 462).

In the early 1980s, the sociologist Niklas Luhmann (1995), appropriated and generalised Maturana and Varela's notion of autopoiesis to describe the autonomous and self-referential operations of social systems, which he elucidated in his book, titled *Social Systems* (Rasch and Knodt 1994). Luhmann was one of Habermas's most prominent critics, opposing his view of universal principles in favour of principles that are self-referential, and therefore paradoxically based on themselves (Arnoldi 2001). Poli (2009) explains Luhmann's application of the theory of autopoiesis to social systems as follows: for social systems to be seen as self-referential systems, it is important to acknowledge a degree of systemic stability. Luhmann follows Talcott Parsons (1951) in arguing that the reproduction of social systems is contingent on the reproduction of their (social) roles (i.e. patterns of action typical of a specific system). Luhmann however goes further than Parsons in providing a firmer basis for social roles in terms of meaning: 'the reproduction of a social system' he argues, 'is grounded on the reproduction of meaning, e.g. through education and other socializing functions' (Poli 2009: 9).

Luhmann (1986) further explains that the units of meaning used by a social system for its reproduction are communications, which he, building on Karl Bühler's (1934) work, views as consisting of information, utterances, and understanding. Understanding 'refers to what the *receiver* grasps from the previous two aspects of a communication' (Poli 2009: 10). Since all three components are necessary to form a communication, communication can never be attributed to a single individual (due to the role of the receiver). As such, Luhmann defines communications as the basis for social acts.

Using this theory, Luhmann divides modern social systems into different *functional systems*, depending on the domains of practice e.g. law, economy, science, art, etc. Each of these subsystems is further evaluated in terms of relevant and irrelevant communications (which is a distinction valid to all subsystems), as well as in terms of a function-specific distinction. This function-specific distinction is unique to the given domain, for example, the distinction between legal or illegal actions applies to the legal system, whereas the distinction between true and false propositions/theories applies to the scientific system. Luhmann also draws distinctions between social systems based on specific subtypes (such as interaction or organisation). These subtypes are based on types of communications, for example face-to-face communication is applicable to interpersonal interactions, whereas decisions are applicable to organisations.

What is of note for this discussion is that '[a]ll the communication takes place within the system; there is no communicative exchange between the system and its environment' (11). Although not thermodynamically-closed, Luhmann maintains that social systems are informationally-closed to their environments. However, social systems are able to reproduce the system/environment distinction (wherein the environment perturbs the system and triggers internal processes) within the system itself. This ability of the system 'to apply to itself the distinction between the system and its environment requires that the system be capable of observing itself' (11). Poli (11) claims that it is this 'observational re-entry' that both 'generates the structure of the system [and] constitutes the second level (or cycle) of autopoietic reproduction'.

The theory of autopoiesis not only had a profound impact on Luhmann's understanding of social systems, but also radically altered the cybernetics paradigm (Hayles 1999). More specifically, it brought about two major shifts: namely, a shift from observed systems to the observer of systems; and a shift from message, signal, and information as that which circulates through systems to the mutually constitutive interactions between the components of a system. As such, Hayles credits Maturana and Varela with introducing important insights regarding reflexivity and the role of the observer (along with the implications that this holds for the notion of scientific objectivity), and for drawing attention to the specificity and concreteness of embodied processes.

However, Hayles criticises the idea of operationally or organisationally-closed systems, and remains unconvinced that Maturana and Varela are able to solve the problem of solipsism. Although Hayles concedes that Maturana and Varela's (1987: 242) statement that: 'We do not see what we do not see, and what we do not see does not exist' cannot be properly understood without contextualising it in terms of the observer's sensory perceptions, she nevertheless questions the consequences that the notion of operational closure holds for our understanding of reality. In this regard, consider the following citation (Hayles 1999: 158):

> But what if "the observer" ceases to be constructed as a generic marker and becomes invested with a specific psychology, including highly idiosyncratic and possibly psychotic tendencies? Will the domains of self-conscious observers fail to stabilize external reality? Will the uncertainties then go beyond questions of epistemology and become questions of ontology? Will the observation that "what we do not see does not exist" sink deep into the structure of reality, undermining not only our ability to know but the ability of the world to be?

Ultimately therefore, the constructivist dimension of operationally or informationally-closed systems carries solipsistic and relativist implications when applied to human and social systems. With regard to social systems, it is noted that Luhmann's work undoubtedly holds important insights for conceptualising these systems as radically immanent systems, but the problems associated with endorsing a strong view of recursivity (where a system maintains itself only in terms of its own operationally-closed processes) ultimately undermine the value of his theory.[2]

[2] See Sect. 2.5 for a discussion on how the principle of organisation in complex systems allows for an opening in informationally and operationally-closed systems.

2.2.3 Third-Order Cybernetics and Artificial Life

As stated in the previous section, Maturana and Varela understand autonomy to be the central feature of the organisation of the living, and it is precisely the concept of *autonomy* that led Varela to the field of artificial life. In the *Proceedings of the First European Conference on Artificial Life*, Varela and his co-author, Bourgine (1992: xi), write that '[a]utonomy in this context refers to [the living's] basic and fundamental capacity to *be,* to assert their existence and bring forth a world that is significant and pertinent without being predigested in advance.' What differentiates third-order cybernetics from second-order cybernetics is the emphasis on the capacity to bring forth a world through self-organisation (defined as the development of systemic structures independently of external influences) and emergence (defined as the appearance of phenomena that depend on, but cannot be fully explained in terms of, a base). Here again, we see a subtle shift in how the characteristic of self-organisation functions in cybernetic theory: whereas in second order cybernetics, self-organisation is associated with homeostasis (or system's maintenance), in third-order cybernetics, self-organisation refers to a system's ability to 'evolve in unpredictable and often highly complex ways through emergent processes' (Hayles 1994: 463).

Emergence is also the quality that distinguishes artificial intelligence (AI) from artificial life (AL). Hayles (1999: 239) succinctly explains as follows: 'Whereas AI dreamed of creating consciousness inside a machine, AL sees human consciousness, understood as an epiphenomenon, perching on top of the machinelike functions that distributed systems carry out.' One of the most famous examples of artificial life software in von Neumann's self-reproducing cellular automata. As early as 1949, von Neumann was already observing a phenomenon for which he had no proper name, but which he later referred to as *complexity* (Rasch 1991). This complex phenomenon that von Neumann was observing was the ability of simple automata to produce complex structures through processes of self-organisation and emergence.

With regard to these cellular automata, one sees how extremely complex patterns emerge from initially undifferentiated automaton states, which are presented as pixels on a computer screen (see Fig. 2.2). These automata function as finite state machines that follow simple rules such as 'on if two neighbours are on, otherwise off'. The automata's states are continuously updated as they move through successive generations. On a computer screen, these automata give the impression of being alive, which has led some researchers to view them as a model for life (Hayles 1999). In this regard, Christopher Langton (1995: xi)—who supports a view of strong a-life—writes that not only will artificial life teach us much about reality, but it 'will ultimately reach beyond biology… [to] include culture and technology in an extended view of nature.'

Although von Neumann's work predates third-order cybernetics, it is definitely a forerunner to the simulated worlds of virtual reality and artificial life that have entered both professional and consumer markets today, and that have created the hope of transforming consciousness into an informational pattern (see Moravec 1988).

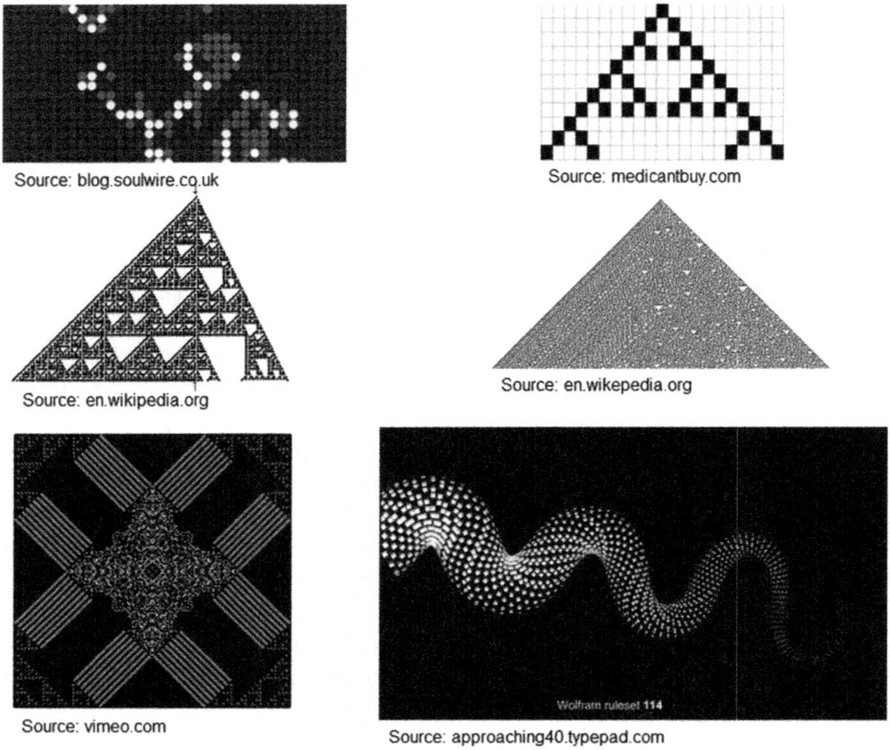

Fig. 2.2 'Cellular automata'

A prominent question that baffled von Neumann and subsequent researchers, and which is still relevant today, is how high-level computations can emerge from the underlying structure of the cellular automata (Hayles 1999). To explain this phenomenon, von Neumann postulated a *complexity barrier*, which is the point past which systems experience not only quantitative, but also qualitative changes (Rasch 1991). Von Neumann (1966: 80) offers the following description of the complexity barrier:

> There is thus this completely decisive property of complexity, that there exists a critical size below which the process of synthesis is degenerative, but above which the phenomenon of synthesis, if properly arranged, can become explosive, in other words, where synthesis of automata can proceed in such a manner that each automaton will produce other automata which are more complex and of higher potentialities than itself.

This idea of a complexity barrier was further explored by Christopher Langton, Norman Packard, and Stuart Kauffman. All three theorists postulated that the requisite variety and creative tension needed for emergent behaviour exists in the boundary area between order and chaos (Horgan 1995). John Horgan (106) reports that, in their experiments with cellular automata, Langton and Packard note that 'a system's computational capacity… peaks in a narrow regime between highly periodic and chaotic behavior.' As a result, they coined the popular term *the edge of chaos* (Langton 1992), which is widely used in definitions of complex systems.[3]

2.3 Cybernetics and Philosophical Complexity

At the beginning of the previous section, it was stated that the movements of cybernetics are important in the context of this work because of the influence that they have had on our understanding of philosophical complexity. In this section, this influence will be discussed at the hand of some of the features of philosophical complexity.

Feature 1: Complex systems are characterised by richly-interconnected and organised components.

In focusing attention on the principles of entropy, information, and feedback, first-order cybernetics influenced the development of philosophical complexity by placing the emphasis on the interactions between objects (regardless of their nature), rather than on the objects themselves. Philosophical complexity theorists view complex systems as inherently relational, which means that the interrelations between the components of complex systems constitute the focus of study in this field. Cilliers (1998) makes the following observations regarding the nature of these interactions:

- they can be physical or informational;
- the interactions are fairly rich, 'i.e. any element in the system influences and is influenced by quite a few other ones' (3);
- the interactions have a short range, but these local interactions can have large, non-linear systemic effects. This implies that systems-level order (and transformations) emerge because of interactions amongst components at lower levels of the system;
- they are asymmetrical (which means that certain relations are more heavily-weighted and influential within the system than others); and,
- there are positive (stimulating) and negative (inhibiting) feedback loops in the interactions, which means that no direct link is necessary for distant elements to interact.

Feature 2: Complex systems cannot be fully known.

The emphasis on self-reflexivity during the period of second-order cybernetics—which brought forth a shift from observed systems to the observers of systems—has had a large impact on how philosophical complexity theorists understand their ability to know complex systems. This has led to inquiries into the nature of modelling and the status of models. The basic argument is that, since complex systems cannot be understood in all their complexity, we are required to model, which in turn means that 'complex thought requires the integration of the observer and the conceiver in

[3] See Sect. 2.6.1 for more detailed descriptions of self-organised criticality and the edge of chaos.

its observation and conception' (Morin 2008: 51). In Chuck Dyke's (1988: 5) words: 'Not only are the phenomena to be studied complex, but scientific practice itself is a phenomenon of organized complexity. The complexity of the investigation must be studied along with the complexities investigated.' Following from this, Luhmann (2000: 46) describes the consequence of knowledge generation as a paradox:

> The self-description of the self-transparent system has to use the form of a paradox, a form with infinite burdens of information and it has to look for one or more distinguishable identities that "unfold" the paradox, reduce the amount of needed information, construct redundancies, and transform unconditioned into conditioned knowledge… [but] the question of the unity of the distinction always leads back to the paradox—and one can show this to others and accept it for oneself.

Transforming unconditioned into conditioned knowledge is not only a technical exercise (involving skill and knowledge) but also a normative exercise (to the extent that the observer is implicated in her observations). To understand complex systems (i.e. to create conditioned knowledge) means that a measure of reduction and simplification is unavoidable, even if—like Joyce Carol Oates—we are loath to simplify, and even if we recognise that such simplifications cause distortions of the systems under study (in that we have no direct access to reality).

The issue at stake here however is not so much the paradox itself. Rather, the issue is whether we recognise the paradoxical status of our models or conceptual frames; or, in Luhmann's (46) words, whether 'one can show this to others and accept it for oneself'. This paradoxical status of models amounts to the recognition that 'the progress of knowledge is at the same time the progress of ignorance' (Morin 1992a: 361). This is because the process of creating conditioned knowledge introduces noise into the system, which degrades information and the progress of simple knowledge. This understanding of noise as the outcome of the observer problem (and hence internal to the creation, and communication, of information) is absent from Shanonnian information theory, in which noise is defined solely in terms of 'the external threat to the integrity of information' (309). In other words, in treating information in purely physical terms, Shannon and Weaver disregarded the anthropo-social dimension, which is fundamental to meaning formation and communication.

The main value of second-order cybernetics is that it uncovered the pretense that we have access to noumena as such. Morin argues that the consequence resulting from the loss of the absolute is that we need to establish relations between phenomena that we have previously treated within the paradigm of disjunction. Specifically, we should establish as positive and new the relation between the physical, the biological, and the anthropo-social (or Matter, Man, and Society). As difficult and illogical as the task may seem, contending with this loop and translating the physical and biological into the anthropo-social, is unavoidable for the reason that the observer is integrated into every observation, with the consequence that:

> All knowledge, even the most physical, undergoes a sociological determination. There is in all science, even the most physical, an anthropo-social dimension. *By that very fact, anthropo-social reality projects itself and reinscribes itself at the very heart of physical science* (5).

Although this study is primarily concerned with the anthropo-social dimension, the central insights emerging from Morin's introduction to his threefold epistemological looping (which forms both the motivation for, and basis of, his *Method*) are nevertheless of central importance to any treatment of complexity. In the translator's introduction to *Method*, Roland Bélanger (1992) summarises these insights as follows: we should (1) discard the Cartesian subject-object dichotomy because the anthropo-social dimension is present in all knowledge; (2) at all times recognise the inescapable looping of the physical, the biological, and the anthropo-social; (3) accept that uncertainty is the only point of departure; and, (4) reject simplifications (in terms of idealisations, rationalisations, and standardisations) in favour of a complex or spiral understanding of knowledge. In other words, we should seek to overturn the knowledge paradigms of disjunction and universalism in favour of the paradigm of complexity, which recognises and engages with the observer problem.

Feature 3: Complex systems are structured and self-organising.

Maturana and Varela's emphasis on autopoiesis, as well as Varela's work on AL, influenced philosophical complexity by drawing attention to the way in which complex systems develop through a process of self-organisation, which is characterised by both a temporal and a structural dimension. When the components of systems interact (through means of competitive and cooperative behaviour), dynamic structures emerge over time. Self-organisation is defined as 'a process whereby a system can develop a complex structure from fairly unstructured beginnings' (Cilliers 1998: 12), or whereby 'internal structure can evolve without the intervention of an external designer or the presence of some centralised form of internal control' (89). Self-organisation is partly the outcome of a system's memory or history, in that the past codetermines the present, and the present codetermines the future.

These definitions serve to dispel the popular notion that complex systems are flat systems. Only homogenous or chaotic systems are flat, because the complex processes that lead to the emergence of dynamic structures are absent (Cilliers 2001). The brain is an excellent example of a self-organising system. The brain functions as a neural network. It is a network of chemically-connected or functionally-associated neurons. The interconnections between neurons are called synapses. Over time, certain pathways are established in the brain, meaning that some of the synapses are reinforced through impulses, whereas others die off. In this way, structure develops as groups of neurons are selected, reinforced, and transformed through interaction with their environment. This implies that a fairly undifferentiated brain develops structure or consciousness over time (Cilliers 1998).

Complex systems also contain a number of nested systems. However, one should remain cognisant that a nested system is very much the product of the description that one gives to the system. For example, the brain can be viewed as either a nested system within a larger human system, or it can be defined as a complex system in its own right, depending on one's level of analysis. Nevertheless, regardless of the level of analysis employed, structures are indispensible for systemic development; since, as Cilliers (89) argues, the structure constitutes 'the internal mechanism developed by the system to receive, encode, transform and store information on the one hand, and to react to such information by some form of output on the other.' The nature of systemic structures are however very much contingent on the particularities of the system under study: whereas some structures are more durable (for example, brain pathways), others are more volatile and ephemeral (for example, patterns generated by cellular automata) (Cilliers 2001).

Feature 4: Complex systems exhibit emergent behaviour.

Third-order cybernetics (which focuses on artificial life) has influenced philosophical complexity in showing that there is a relation between self-organisation and emergence—both of which are central to any understanding of complex systems. Self-organisation is a necessary condition for emergence, which is defined as 'the idea that there are properties at a certain level of organization which cannot be predicted from the properties found at lower levels' (Emmeche et al. 1997: 83). Specifically, self-organisation draws attention to the structural and temporal dimensions of emergence. Complex systems must be able to 'learn' from experience and 'remember' past encounters (Cilliers 1998). Cilliers (92) explains that '[i]f more 'previous information' can be stored, the system will be able to make better comparisons. This increase in complexity implies a local reversal of entropy, which necessitates a flow of information through the system.' It is therefore only possible for systems to develop complex structures by processing information, and developing 'memory'. The previous discussion on how the brain develops serves as a good example of self-organisation. However, the mind is also not equivalent to the brain. Although it cannot exist without the brain, the mind is also 'greater' or 'more' than the brain (which is made up of self-organised neurons or synapses). The mind, in other words, is an emergent phenomenon, which means that it is dependent on a base (i.e. the brain), but simultaneously supersedes that base. It therefore seems that, whilst self-organisation is a necessary condition for emergence, it is not a sufficient condition. However, despite this, there is still much debate regarding the nature of emergence (see Bedau and Humphreys 2008).

Feature 5: Complex systems are incompressible due to the workings of a complex notion of causality.

Despite the influence that third-order cybernetics has had on our understanding of emergence, the coupling of self-organisation and emergence to the phenomenon of artificial life (as is the case in third-order cybernetics) has not led to the anticipated advancements in artificially modelling life. From a philosophical complexity perspective, the reason for this is that the field of artificial life is predicated on a restricted understanding of complexity, whereby simple rules are seen to give rise to complex behaviour. From the perspective of philosophical complexity, complexity is viewed as irreducible (in other words, we cannot compress complex systems without discounting some of the complexity). Complexity cannot be traced back to simple constitutive rules for the reason that this scientific methodology does not take sufficient account of the 'countless, local non-linear, non-algorithmic, dynamic interactions, [which]... cannot be described completely in terms of a set of rules' (Cilliers 2000: 46). In emphasising these complex, non-linear relations, philosophical complexity theorists also reject the description of complex systems in terms of *the edge of chaos*, since such a description presupposes a definable point of *self-organised criticality*—the point of complexity, so to speak. As soon as dynamic and complex interactions between systemic components exist, systems start developing structures. Trying to pinpoint optimal levels of organisation, through recourse to terms such as self-organised criticality, again denies a measure of complexity. This is because complexity is not an additive process, but is rather the result of a complex causality, in which:

> a) Like causes can lead to different and/or divergent effects… b) Different causes can produce like effects… c) Small causes can bring about very large effects… d) Great causes can entail very small effects… e) Some causes are followed by the opposite effects… f) The effects of antagonistic causes are uncertain (Morin 1992a: 270).

Morin (270) further defines complex causality as 'circular and interrelational' with the consequence that:

> cause and effect have lost their substantiality: cause has lost its all-powerfulness, effect its all-dependence. They are relativized by and in each other, they are transformed into each other.

Therefore—and as stated in Chap. 1—the argument from a philosophical complexity viewpoint is that if complexity is irreducible, then we cannot wholly master complex systems in terms of a scientific understanding of the trajectory of the development of complex phenomena. Complexity manifests at many different levels in the system, and the relations between components cannot allow for predictive or quantifiable behaviour as such behaviour is necessarily based on a linear understanding of causality.[4]

2.4 General Systems Theory

Like cybernetics, GST has provided important challenges to the traditional view of science, and has greatly enriched our knowledge of the organisation of systems (as a set of elements standing in interaction amongst themselves and with the environment). Morin (2008) however argues that GST has a much wider scope than cybernetics. Indeed, it is quasi-universal to the extent that 'all known reality, from the molecule to the cell to an organism to a society, can be conceived of as systems.'

The biologist, Ludwig von Bertalanffy (1972), provides a good description of the commonalities between systems theory and cybernetics. He remarks that although GST did not spring out of the war effort (as is the case with cybernetics), both movements nevertheless share an interest in the organisational and teleological behaviour of systems. Von Bertalanffy also notes that, despite the fact that cyberneticians and systems theorists have different starting points (technology versus science—especially biology) and use different basic models (feedback circuits versus dynamic systems of interaction), both approaches present a challenge to the mechanistic conception of the universe. On this point, GST shares the cybernetics agenda, which is defined by Frank et al. (in von Bertalanffy 1972: 414) as 'the search for new

[4]Also see Sect. 2.6.2.

approaches, for new and more comprehensive concepts, and for methods capable of dealing with the large wholes of organisms and personalities'.

Von Bertalanffy however is also quick to remind us that the systems approach is not new to the history of ideas. One formulation of the basic systems problem, which still remains valid today, is Aristotle's statement that the whole is more than the sum of the parts. Other systems thinkers that von Bertalanffy mentions are Dionysius (who introduced the term *hierarchic order*), Nicholas of Cusa (who introduced the notion of *coincidentia oppositorum*, which refers to the fight amongst components of a system that nevertheless form a unified whole), Leibniz (whose hierarchy of monads closely resembles modern systems), and Hegel and Marx (whose works on dialectics shed light on the interactions in systems). As such, von Bertalanffy notes that:

> the problems with which we are nowadays concerned under the term "system" were not "born yesterday" out of current questions of mathematics, science, and technology. Rather, they are a contemporary expression of perennial problems which have been recognized for centuries and discussed in the language available at the time (408).

Modern day systems theory developed from the recognition that the paradigm of classic science, wherein phenomena are studied in terms of isolable elements, was no longer adequate for explaining complex matters. Rather, and as noted by Morin (1992a), increasing attention was given to the interrelations of elements that give rise to a global unity, and one field in which this new insight was readily appropriated was that of biology. Already in the 1920s, von Bertalanffy (1972: 410) had identified the organisation of organisms as 'the fundamental character of the living thing.'

In order to capture this idea of organisation, von Bertalanffy (410) coined the terms *organismic biology* and *the system theory of the organism*. This organismic programme was the germ for what later became known as GST. Von Bertalanffy states that he first formulated the notion of GST orally in the 1930s, and then in various publications after the Second World War. An early formulation of GST reads as follows:

> General Systems Theory is a logico-mathematical field whose task is the formulation and derivation of those general principles that are applicable to "systems" in general. In this way, exact formulations of terms such as wholeness and sum, differentiation, progressive mechanization, centralization, hierarchical order, finality and equifinality, etc., become possible, terms which occur in all sciences dealing with "systems" and imply their logical homology (411).

The notion of organisation upon which the principles of GST are based, pertains to the 'set of elements of a system standing in interrelation among themselves and with the environment' (417). Organisation is thus the term that ties systemic interrelations with a systemic totality (Morin 1992a), and the system with the environment.

Von Bertalanffy defines the system's relation with its environment as open. This means that systems exchange energy and matter with their environment,[5] which,

[5] Open systems are juxtaposed with closed systems that are defined by an exchange of energy only (for example, the earth that receives radiation from the sun), as well as with isolated systems in which no energy/matter exchanges take place (for example, a can of soup) (Morin 1992a).

according to von Bertalanffy, is the hallmark of living systems. The idea of open systems originated as a thermodynamic concept (Morin 2008). In thermodynamics, system's equilibrium equates to system's death as a constant source of energy is needed for systemic maintenance. However, if systems are completely random, then they also have no capacity for complex behaviour. For living systems to function, they must therefore be in a state of nonequilibrium or energetic flux, but must simultaneously retain a measure of stability and continuity.

Ilya Prigogine, the Belgian chemist and prominent complexity theorist, demonstrated the principle of nonequilibrium in his work on dissipative structures for which he received a Nobel Prize. Dissipative structures are '"pumped" chemical cells that never achieve equilibrium but oscillate between multiple states' (Horgan 1995: 108). Prigogine and Stengers (1984: 143) state that these dissipative states—or the formation of new dynamic states of matter—illustrate '[t]he interaction of a system with the outside world, [and] its embedding in nonequilibrium conditions'. These nonequilibrium conditions further show that '[d]isorder does not simply destroy order, structure and organization but is also a condition of their formation and reformation' (Taylor 2001: 121). This insight is not recognised in Shannon's understanding of noise in communication systems.

2.5 General Systems Theory and Philosophical Complexity

As with cybernetics, GST has had a big influence on the development of philosophical complexity—particularly in terms of the focus on both a system's relationship with its environment, and systemic organisation. Morin however treats both of these focus areas in a manner that extends the insights gleaned from systems theory. In this section, these insights are discussed at the hand of additional features of complex systems.

Feature 6: Complex systems are open systems, as well as operationally-closed and bounded systems.[6]

As mentioned above, a central strength of GST is that the focus is on the relation between complex systems and their environments. However, before exploring the nature of open systems in more detail, it is firstly necessary to determine whether there is a productive way in which to reconcile the notions of operationally-closed and open systems.

To recall: Maturana and Varela's theory of autopoiesis in second-order cybernetics de-emphasises this classical distinction between open and closed systems in favour of an understanding of operational-closure, which denotes 'the generation of structure, understood as the set of constraints governing the system's internal processes' (Poli 2009: 8). The idea of operational-closure need not however be at odds with the idea of systemic openness. Morin criticises the rigid opposition between open and closed systems, arguing that complex systems are neither completely

[6]Features 6, 7, and 8 will be revisited in Chap. 7 (especially Sect. 7.2).

closed, nor completely open. Morin's elucidates as follows on why the principle of organisation necessitates both systemic openness in autopoietic systems and systemic closure in open systems:

> active organizations of systems called open insure the exchanges, the transformations, which nourish and effect their own survival: the opening allows them to ceaselessly form and reform themselves; they are reformed by closing, by multiple loops, negative retroactions, recursive uninterrupted cycles. Thus the paradox imposes itself: an open system is opened in order to be closed, but is closed in order to be opened, and is closed once again by opening. The closing of an "open system" is the loop on itself… it is active reclosing which insures active opening, which insures its own closing… and this process is fundamentally organizational. Thus, living organization is opened in order to be closed again (to insure its autonomy, to preserve its complexity) and is closed again in order to be opened (to exchange, communicate, enjoy, exist…) (Morin 1992a: 133).

Morin thus foregoes the opposition between organisational openness and closure, in favour of a view in which openness and closure are shown to mutually coproduce complex systems. The paradoxical relationship between systemic openness and closure however remains a difficult issue to grasp conceptually. In Chap. 3, this issue will be explored at length at the hand of the notions of a restricted (closed) economy and a general (open) economy. Although this discussion proceeds predominantly at the hand of the philosophical literature on economy, the analysis is nevertheless helpful in showing why operational closure cannot be complete.

At this juncture, it is however worthwhile to note that it is difficult to study these open-closed (i.e. complex) systems. Morin (2008: 11) contends that part of the difficulty lies in the fact that '[t]he environment is at the same time intimate and foreign: it is part of the system while remaining exterior to it.' Even though we can only know the environment in terms of the system, the environment should not be reduced to a feature of the system itself, as Luhmann would have it. Rather the environment should be viewed as something that stands apart from the system, but nevertheless interacts with the system in order to *constitute* (as opposed to merely maintain) the system. In other words, and as noted by systems theorists, a system and its environment should be treated as both real, physical properties, and mental categories or ideal models (Morin 1992b: 379).

Moreover, Cilliers (2001) argues that in open systems, the boundary between the system and its environment acts as the interface that participates in constituting the system as different from the environment. He further states that in critically-organised systems we are never far from the boundary, since there is always a short route from any richly inter-connected component to the 'outside' of the system. This problematises the inside-outside distinction, as the boundary of a critically-organised system is folded in.

Feature 7: Complex systems are thermodynamically and organisationally-open.

Having established that systems are both open and operationally-closed, it is now necessary to determine in which way(s) complex systems are open. In GST, systems are defined as thermodynamically open (i.e. they exchange energy and matter with their environments), with the consequence that the laws of the organisation of the living are laws of nonequilibrium or stabilised dynamics.

However, Morin (1992a) contends that systems are also open in an organisationist sense. This means that the activity of the system (in relation to the environment) is what constitutes the system as open. Such activity is not only characterised by order, but also by disorder. In fact, Morin (130) argues that '[t]he more [the system] becomes complex, the more its order is mixed more and more intimately with disorders, the more antagonisms, uninhibitions, fortuities play their role in the being of the system and its organization.' Thus organisational openness has two faces: the first is an 'organisationist expression' (which includes disorder or anti-organisation), whereas the second concerns a 'thermodynamic expression' (which introduces entropy)[7] (130). Both the internal disorder of the system and the entropic forces acting upon the system, mean that disorganisation and reorganisation are as characteristic of systems as organisation itself.

Feature 8: Organisation is the key feature linking the interrelations between systemic components to the emergence of a system, defined as a unitas multiplex.

For Morin, the principle of organisation is the key to understanding complex systems. This principle provides the link between systemic interrelations and the emergence of a global structure, endowed with qualities unknown at the level of systemic components. Furthermore, this principle is itself the product of 'chance encounters, in the copulation of disorder and order in and by catastrophe' (100). As such, Morin identifies a loop of mutual co-production between 'order → disorder → interaction → organization' (49). Yet, he argues that even though original constraints and potentialities give rise to both order and disorder; once established, '[o]rder and organization… are capable of gaining ground on disorder' (49; italicised in the original). This fact allows for the emergence of a second loop of mutual co-production, in which the interrelations between elements give rise not only to organisation, but also to systems; or, in the words of Morin 'the upsurge of interrelation, of organization, or system, are three aspects of the same phenomenon' (100).

From the above, we conclude that the coupling of order and disorder is necessary for, and antagonistic to, organisation. This point is not sufficiently recognised in GST, with the consequence that insufficient attention is given to the paradoxical manner in which systems emerge. As a consequence, much of the work in this field is characterised by the problem of reductionism. However, unlike the traditional scientific approach—in which the basic constituting elements are studied in order to gain knowledge of a composite (Morin 2007)[8]—systems theorists tend to simplify and reduce the constituting elements to the composite. More specifically, systems theorists are often guilty of employing the principle of holism, which Morin (1992b: 372) describes as follows:

> Holism is a partial, one-dimensional, and simplifying vision of the whole. It reduces all other systems-related ideas to the idea of the totality, whereas it should be a question of confluence. Holism thus arises from the paradigm of simplification (or reduction of the complex to a master-concept or master-category).

[7] The organisational character of opening will be discussed in more detail in Chap. 7.

[8] See Sect. 2.6.

Instead of conceiving of systems in terms of a global unity (as is the case in GST), Morin argues that we should view systems and their component parts in terms of the principle of organisation. This gives rise to a '*unitas multiplex*' (373), which is the outcome of both the relation between order and disorder and the coupling of antagonistic terms.

Feature 9: The component parts of complex systems have a double identity.

The components in complex systems remain antagonistic to the extent that they retain their own individual identities that cannot be reduced to one another or to the whole. Consider the example of Maya, who works as an academic. Maya's identity is constituted by more than merely her role as an academic within the system of the university, even though she cannot live out her larger identity in terms of the university. Hence, a part of her identity cannot be explained in terms of the system of the university. At the same time however the coupling of the parts implies a common identity, which constitutes their citizenship in the system. The fact that Maya, for example, interacts with other academics and students on a professional basis, constitutes behaviour that supports the goals of the university, and thus confirms her identity as an academic in the system of the university. In other words, the systemic components have a double identity. The 'system is not only a composition of unity out of diversity, but also a composition of internal diversity out of unity' (373). When thinking about systems, this double identity needs to be accounted for, because if we forego the diversity principle, our thinking becomes increasingly homogenised (holism); but if we forego the unity principle, our 'thinking becomes a mere catalogue and loses unity' (373).

Feature 10: Complex systems are non-additive due to the manner in which the parts and the whole are related.

Taking cognisance of this double identity is not enough: Morin (1992b) states that we should also account for the complex character of these interrelations or the fact that complex systems are non-additive. This, firstly, means that we should respect the age-old truism that 'the whole is greater than the sum of the parts' in that systemic attributes cannot be reduced to the parts alone, but are the result of interconnections between the parts. Secondly, and less widely recognised, is the fact that the whole is also less than the sum of its parts, since some of the qualities of the parts are suppressed under the constraints that result from systemic organisation. The previous example pertains: within the system of the university, Maya cannot exercise her role as a mother since this role is inappropriate within the university. Therefore, although the fact that Maya is an academic forms part of her larger identity, this role which manifests in the system of the university, does not exhaust her identity. Hence, the system of the university is less than the sum of the individuals working in the university. Thirdly, and perhaps most counter-intuitively, the whole is also greater than the whole, due to the dynamic organisation that takes place in systems where local interactions between components give rise to emergent properties. For example, the unit of the family or society as such has an identity that cannot be adequately explained merely in terms of the relations between family members or the members of society. The previous example of the brain being 'more' than the mind also pertains.

Feature 11: Complex systems are defined by organisational recursion, which implies that the identity of components and systems are coterminous.

In contrast to reductionism and holism, complexity thinking requires that one must try to comprehend the relation between the whole and the parts. What is important here is the *relation* itself: knowledge of the whole is not enough, and knowledge of the parts is not enough. One must substitute the principle of reductionism with a principle that conceives of whole-part mutual interaction (Morin 2007). These mutual interactions result in, what Morin (2008: 49) terms, 'organizational recursion', which means that relational components create, engage in, and challenge the systems to which they belong; and the systems simultaneously serve to shape the components. Morin (50) uses the idea of the hologram to explain this latter point: '[i]n a physical hologram, the smallest point of the hologram image contains the quasi-totality of information of the represented object. [Therefore] not only is the part in the whole, but the whole is also in the part.' The example of society again pertains: not only do individuals produce society through their interactions, but from early childhood, society enters us through a process of socialisation that supplies us with language and culture. An implication of organisational recursion is that the identity of systemic components and the systems themselves are coterminous, i.e. they arise and die together.

2.6 Philosophical Complexity and Complexity Science, or the Distinction Between Complexity and Complicatedness

A cursory glance over the above critical comparisons between philosophical complexity, cybernetics, and GST reveals that these latter fields diverge from philosophical complexity with respect to their more conservative elements, specifically the implicit endorsement of reductionism. In first-order cybernetics all of life is reduced to the machine concept, in which control is the defining feature; in second-order cybernetics, the environment is reduced to the system, with potential solipsistic consequences; in third-order cybernetics, complex behaviour is reduced to simple rules, which give rise to an additive view of complexity; and in GST, systemic components are reduced to the whole, resulting in the view of a system as a global unity. The endorsement of reductionism also defines much of the work undertaken in the field of scientific complexity and, in this section, important historical developments in complexity science that emerged from the work done at the Santa Fe Institute are (very!) briefly noted, with the goal of substantiating this latter point.

2.6.1 Complexity Science

John Holland (1993), a member of the executive committee of the board of trustees at the Santa Fe Institute,[9] noted that—despite appearances—a range of complex systems 'do share significant characteristics, so much so that we group them under a single classification at the Santa Fe Institute, calling them complex adaptive systems [CAS]' (1). Holland (1) further states that CAS 'signals [the Institute's] intuition that there are general principles that govern all complex behavior, principles that point to ways of solving… [our] problems.' Not much has changed since this formulation, and the explicit mission of the Institute is described as follows on its website:

> to discover, comprehend, and communicate *the common fundamental principles* in complex physical, computational, biological, and social systems that underlie many of the most profound problems facing science and society today [my italics].[10]

Some of the more prominent attempts at formulating the principles of complexity include Langton's (1992) ideas on life at *the edge of chaos*, Per Bak's (1997) work on *self-organised criticality*, and Kauffman's (1993) work on *attractors* and *strange attractors*. As noted earlier in the analysis,[11] Langton studied the phase transitions of the behaviour of cellular automata, and noticed that complex, almost life-like behaviour exhibits on the border between order and chaos. Although it is impossible to determine the exact edge, Langton was able to conclude that the lambda parameter (i.e. a function containing expressions) lies somewhere between 0 and 1. If the lambda value lies too close to 0 then the behaviour of the cellular automata become too ordered, and if the lambda value lies too close to 1 then the behaviour becomes too chaotic to be considered complex (Eck n.d.). Like Langton, Kauffman and Bak also tried to isolate the moment of complexity. Kauffman's (1993) work on attractors and strange attractors is focused on determining the numerical values towards which sequential dynamical systems or Boolean networks that are randomly selected from the set of all available Boolean networks of a particular size tend to evolve. Bak studied the self-organising behaviour of cellular automata in order to determine the moment of self-organised criticality. He uses the sand pile as a visualisation for self-organised criticality: once the system is critically-organised, adding any further grains of sand will cause an avalanche. He found that although the timing and size of avalanches were unpredictable, the distribution of avalanches and their timing displayed an interesting regularity (Teran 2001). The findings yielded by the above-mentioned experiments led complexity scientists to conclude that it should be possible to pinpoint the moment at which dynamic systems begin to exhibit complex behaviour, as well as to define the set of interactions that lead to complex behaviour.

---

[9] http://www.santafe.edu/about/people/profile/John%20H.%20Holland Downloaded 16 March 2015.

[10] http://www.santafe.edu/about/mission-and-vision/ Downloaded 16 March 2015.

[11] See Sect. 2.2.3.

Despite the substantial contributions that these theorists have made to the theoretical debate, and to work on the formalisation and modelling of complex systems, attempts at viewing these formalisations as models of life, or of constructing unifying theories, have been repeatedly criticised. In this regard complexity theorists have encountered numerous problems, beginning with the term itself. In the words of Horgan (1995: 105): 'Complexologists have struggled to distinguish their field from a closely related pop-science movement, chaos.' The phenomenon of chaos is often described with reference to the butterfly effect, which encapsulates the idea that a relatively minor event could potentially have a far-reaching ripple effect on subsequent historical events. This idea was first introduced by Ray Bradbury in a short story on time-travel, titled *A Sound of Thunder* (1952), but was later popularised by one of the pioneers of chaos theory, Edward Lorenz (Horgan 1995).

Chaos theory describes systems that display non-linear dynamics, bifurcation, a sensitivity to initial conditions, and other mathematically-defined behaviour (Horgan 1995). These characteristics constitute the general principles of chaos theory, but despite the initial excitement over this new theory, 'chaos turned out to refer to a restricted set of phenomena that evolve in predictably unpredictable ways' (105–106). This is because, despite the large-scale effects that fluctuations in the initial conditions of a system may give rise to, once established, the trajectory of chaotic events can be predicted on the basis of deterministic equations. This prompted another pioneer in the field, David Ruelle (in Horgan 1995: 109) to state that 'in spite of frequent triumphant announcements of 'novel' breakthroughs, [chaos] has had a declining output of interesting discoveries.' Although chaos theory did not achieve the success envisioned, chaos theorists did achieve something that complexity theorists at the Santa Fe Institute long hoped for: a precise description of the systems under observation.

Jack Cowan (in Horgan 1995: 104), one of the Institute's founders, notes that the major discovery to have emerged from the Santa Fe Institute is that 'it's very hard to do science on complex systems.' Dominique Chu et al. (2003: 20–21) state that although there is no litmus test for determining what counts as a scientific theory or not, classic scientific theories necessarily conform to three aims, namely:

- Prediction of the future behavior of a system given a set of observational data about it (predictive component);
- Theoretical understanding and/or a description of a system (explanatory component);
- Provision of guidelines and control mechanisms for the intervention and manipulation of systems (control component).

The traditional scientific paradigm however does not hold up when it comes to complex phenomena. There is no 'magic criterion' (Landauer in Horgan 1995: 105) by which to unravel the complexities of nature. Indeed, Horgan (1995: 107) states that the entire field of complexity and artificial life is based on the following seductive syllogism:

> [Premise 1:] There are simple sets of mathematical rules that when followed by a computer give rise to extremely complicated patterns.

> [Premise 2:] The world also contains many extremely complicated patterns.
> Conclusion: Simple rules underlie many extremely complicated phenomena in the world. With the help of powerful computers scientists can root those rules out.

In order to provide further support for this view, consider Holland's (1998: 24–26) approach to complexity, which is based on formal models consisting of 'atomistic building blocks… whose interactions are determined by a set of formal production rules'. This, according to Cilliers (2000), suggests that formal rules are fundamental to complex systems. Cilliers criticises Holland's view of complexity, arguing that something that can be fully understood in terms of a set of rules cannot be rightfully described as complex because the irreducible nature of complexity is not accounted for. Cilliers thus concludes that although we cannot avoid using rules, formal rule-based systems (such as described by Holland) cannot fully capture complexity.

2.6.2 Complicated Versus Complex Systems

Given the above, it is my contention that—on certain points—scientific complexity, cybernetics and GST treat complex systems as complicated systems. Poli (2013: 142) defines a complicated problem as 'originat[ing] from causes that can be individually distinguished; they can be addressed piece-by-piece; for each input to the system there is a proportionate output; the relevant system can be controlled and the problems they present admit permanent solutions.' In other words, complicated systems can be studied via the reductionist methodology. In a footnote, Poli (142) further adds that 'the "complicated" perspective point tends to work with closed systems, while the "complex" perspective point works with open systems'. Poli (142) also provides a list of characteristics that define complex systems, which resonates with the characteristics that have emerged from the discussion thus far:

> complex problems and systems result from networks of multiple interacting causes that cannot be individually distinguished; must be addressed as entire systems, that is they cannot be addressed in a piecemeal way; they are such that small inputs may result in disproportionate effects; the problems they present cannot be solved once and for ever, but require to be systematically managed and any intervention merges into new problems as a result of the interventions dealing with them; and the relevant systems cannot be controlled…

Aside from Poli, Cilliers (1998) also elaborates on the differences between complicated and complex systems. For Cilliers, the central difference is that whereas a complicated system may initially look complex (due to the large number of components that constitute the system, and/or the sophistication of the tasks that the systems can perform), the hallmark of a complicated system is that it is—in principle—solvable (thus affirming Poli's view). An example that has often been used by Cilliers to illustrate the difference between these two types of systems is that of a jumbo jet and mayonnaise: whereas the former is complicated, the latter—as can be attested to by all chefs—is complex!

Given enough information and resources, the dynamics of a complicated system can be fully understood. Poli (2013) however also notes that although complicated systems are in principle solvable, it may not be feasible to build complete models of complicated systems. This is because it may be either too costly or we may not, at the time of modelling, have all the necessary information at our disposal. Models of complicated systems are therefore 'connected to the amount of available data or knowledge', whereas models of complex systems 'are always incomplete and diverge over time' (144).

The reason for why models of complex systems are always incomplete, is because these systems display organised complexity, which, as explained by Weaver (1948), means that systems are inherently complex due to their organising processes.[12] Since only certain aspects of complex systems can be understood at a given time (depending on how we model the system), it also means that complex systems are incompressible. In this regard, Robert Rosen (1985: 424) argues that a system is complex precisely 'to the extent that it admits non-equivalent encodings; encodings which cannot be reduced to one another.'

Poli (2013) notes that despite the fact that the defining difference between a complicated and a complex system is one of type, and not of degree—i.e. '[t]he properties used to classify a system as complicated are different from the properties used to understand a system as complex' (143)—there is a tendency amongst academics and decision-makers to treat complex systems as complicated systems. This tendency—even within the field of complexity science itself—recalls Morin's distinction between restricted and general complexity. Briefly put, under the restricted, scientific view, complexity theorists believe that many aspects of our complex reality are guided by simple patterns that can be discovered. In contrast, those who follow the paradigm of general, philosophical complexity believe that complexity is irreducible and incompressible. Whereas the goal of scientific complexity theorists is to uncover the laws and rules of our complex realities and to develop mathematical formalisms to describe complex behaviour, the goal of philosophical complexity theorists is to advance frameworks that are sensitive to complexity as such.

Morin (2007: 9) argues that the restricted view of complexity spread due to the fact that the term *complexity* could be used to encompass 'ideas of chaos, fractals, disorder, and uncertainty'. Yet, Morin (10) views this as a faulty classification for the reason that 'complexity is restricted to systems which can be considered complex because empirically they are presented in a multiplicity of interrelated processes, independently and retroactively associated.' In other words, these systems are presented and treated as complicated systems, in that the explanatory principle is the principle of reduction, which is supported by the principle of disjunction and the principle of universal determinism. The principle of disjunction consists in separating cognitive difficulties from one another, whereas the principle of universal determinism is the idea that deterministic principles govern the course of cosmic events, past and future. As such, and within this approach, the search for what Morin (10) calls *the laws of complexity* still amounts to the attempt to attach 'complexity

[12] See also Sects. 2.3 and 2.5.

as a kind of wagon behind the truth locomotive, that which produces laws.' Morin concludes that within the restricted view, 'complexity is never questioned nor thought epistemologically'; or, otherwise put, 'one recognizes complexity, but by decomplexifying it' (10).

Morin argues that although restricted complexity rejects general or (what I have termed) philosophical complexity, often dubbing it as 'pure chattering, pure philosophy' (27), he—like Poli and Cilliers—is of the opinion that restricted complexity is no complexity at all. In rejecting 'the epistemological and paradigmatic revolution which complexity obliges' (27–28), restricted complexity evades the very problematic that complexity presents us with, and that has 'invaded all our horizons' (28). As with Morin, Poli (2013: 145) also argues that—contrary to conventional beliefs—'[c]omplex systems are the usual normal case' whereas it is 'complicated systems… [that] are highly distinctive, very special, and therefore rare.' Furthermore, Poli warns that treating complex problems as though they are complicated problems often serves to worsen these problems and further highlights the reductionist tendency to view the type-difference between complicated and complex systems as merely apparent, rather than unbreachable.

The clear message emerging from Poli's and Cilliers' discussion on the difference between complicated and complex systems, and Morin's' discussion on restricted versus general complexity, is that '[t]he real problem… is not to reduce the complication of developments to rules with a simple base, [but to recognise that] *[c]omplexity is the base'* (Morin 1992a: 386). When it comes to complex phenomena we are obliged to abandon our reductivist tendencies, 'to detrivialize knowledge and our worldview' (Morin 2007: 17), and to recognise that the best we can do with complex systems is to learn to 'dance with them' (Meadows in Poli 2013).

2.7 Conclusion

The purpose of this chapter was to present an overview of the history and development of complexity ideas. In the following chapters these ideas will be applied to a specific set of problematics that pertain to the themes of economy, knowledge, identity, and ethics. As stated in the general introduction, these themes will not only be explored at the hand of the ideas introduced in this chapter, but will also be interrogated at the hand of post-structural positions that are sensitive to complexity. The reason for doing so is that I believe that these philosophical insights can deepen our understanding of complexity, specifically as concerns our phenomenological engagement with complexity.

This chapter was largely informed by Morin's vision of complexity. As should be clear at this juncture, the principle of organisation is central to his understanding of how complex systems develop and function, and how complex systems are related to their environments. Morin's understanding of this principle will be returned to in the final chapter. In the following chapter, the nature of complex systems and the organised relations between a system and its environment will be further explored from a philosophical perspective. This will be done at the hand of the philosophical notion of economy, which is defined as a system that is 'dependent upon the limits or constraints determined by the relationships between the components in the system' (Human and Cilliers 2013: 27).

Although the analyses straddle two language games, namely the complex and the philosophical, the insights emerging from these two bodies of knowledge share a common view of the nature of complexity, which greatly eases the difficulty of this transdisciplinary project. Since Morin's views have featured prominently in this chapter, it seems appropriate to end with his summary on complexity, the sentiment of which will be carried forward in the analyses to come:

> Complexity asserts itself first of all as an impossibility to simplify; it arises where complex unity produces its emergences, where distinctions and clarities in identities are lost, where disorder and uncertainty disturb phenomena, where the subject/observer surprises his own face in the object of his observation, where antimonies makes the course of reasoning go astray… (Morin 1992a: 386).

References

Arnoldi, J. 2001. Niklas Luhmann: An introduction. *Theory, Culture & Society* 18(1): 1–13.

Bak, P. 1997. *How nature works*. New York: Oxford University Press.

Bedau, M.A., and P. Humphreys (ed.). 2008. *Emergence: Contemporary readings in philosophy and science*. Cambridge, MA: MIT Press.

Bradbury, R. 1952. A sound of thunder. *Colliers Magazine*, June 28.

Bühler, K. 1934. *Sprachtheorie: Die Darstellungsfunktion der Sprache*. Jena: Fischer.

Byrne, D., and G. Callaghan. 2014. *Complexity theory and social sciences: The state of the art*. Oxon: Routledge.

Cilliers, P. 1998. *Complexity and postmodernism: Understanding complex systems*. London: Routledge.

———. 2000. Rules and complex systems. *Emergence* 2(3): 40–50.

———. 2001. Boundaries, hierarchies and networks in complex systems. *International Journal of Innovation Management* 5(2): 135–147.

Chu, D., R. Strand, and R. Fjelland. 2003. Theories of complexity: Common denominators of complex systems. *Wiley Periodicals* 8(3): 19–30.

Dyke, C. 1988. *The evolutionary dynamics of complex systems: A study in biosocial complexity*. New York: Oxford University Press.

Eck, D.J. n.d. Introduction to 'The Edge of Chaos'. http://math.hws.edu/xJava/CA/EdgeOfChaos.html. Accessed 17 March 2014.

Haraway, D. 1985. A manifesto for cyborgs: Science, technology, and socialist feminism in the 1980s. *Socialist Review* 80: 65–108.

Hayles, N.K. 1994. Boundary disputes: Homeostasis, reflexivity, and the foundations of cybernetics. *Configurations* 2(3): 441–467.

———. 1999. *How we became posthuman: Virtual bodies in cybernetics, literature, and informatica*. Chicago/London: University of Chicago Press.

Holland, J.H. 1993. Echoing emergence: Objectives, rough definitions, and speculations for echo-class models. Working Paper, 93-04-023, Santa Fe Institute.

———. 1998. *Emergence: From chaos to order*. Reading, MA: Addison-Wesley (Helix Books).

Horgan, J. 1995. From complexity to perplexity. *Scientific America,* June 1995: 104–109.

Human, O., and P. Cilliers. 2013. Towards an economy of complexity: Derrida, Morin and Bataille. *Theory Culture Society* 30(5):24–44.

Kauffman, S.A. 1993. *Origins of order: Self-organization and selection in evolution, technical monograph*. New York: Oxford University Press.
Lafontaine, C. 2007. The cybernetics matrix of 'French Theory'. *Theory, Culture & Society* 24(5): 27–46.
Langton, C.G. 1992. Life at the edge of chaos. In *Artificial life II: Santa Fe Institute studies in the sciences of complexity*, vol. 10, ed. C.G. Langton, J.D. Farmer, S. Rasmussen, and C. Taylor, 41–91. Reading, MA: Addison-Wesley.
Langton, C.G. (ed.). 1995. *Artificial life: An overview*, ix–xi. Massachusetts: Massachusetts Institute of Technology.
Lettvin, J.L., H.R. Maturana, W.S. McCulloch, and W.H. Pitts. 1959. What the frog's eye tells the frog's brain. *Proceedings of the Institute for Radio Engineers* 47(11): 1940–1951.
Luhmann, N. 1986. The autopoiesis of social systems. In *Sociocybernetic paradoxes: Observation, control and evolution of self-steering systems*, ed. F. Geyer and J. Van Der Zouwen, 172–192. London: Sage.
———. 1995. *Social systems*. Trans. J. Bednarz, Jr. (with D. Baecker). Stanford: Stanford University Press.
———. 2000. Why does society describe itself as postmodern? In *Observing complexity: systems theory and postmodernity*, ed. W. Rasch and C. Wolfe, 35–49. Minneapolis/London: University of Minnesota Press.
Maturana, H.R., G. Uribe, and S. Frenk. 1968. A biological theory of relativistic color coding in the primate retina. *Archivos de Biologia y Medicina Experimentales, Supplemento No. 1*. Santiago: Chile.
Maturana, H.R., and F.J. Varela. 1980. *Autopoiesis and cognition: The realization of the living*. Dordrecht: D. Reidel.
———. 1987. *The tree of knowledge: The biological roots of human understanding*. Boston: New Science Library.
Moravec, H. 1988. *Mind children: The future of robot and human intelligence*. Cambridge, MA/London: Harvard University Press.
Morgan, G. 2006. *Images of organization*. London: Sage.
Morin, E. 1992a. *Method: Towards a study of humankind*, vol. 1, *The nature of nature*. Trans. and intro. J.L.R. Bélanger. New York: Peter Lang.
———. 1992b. From the concept of a system to the paradigm of complexity. Trans. S. Kelly. *Journal of Social and Evolutionary Systems* 15(4): 371–385.
———. 2007. Restricted complexity, general complexity. Trans. C. Gershenson. In *Worldviews, science and us: Philosophy and complexity*, ed. C. Gershenson, D. Aerts, and B. Edmonds, 5–29. Singapore: World Scientific.
———. 2008. *On complexity*. Trans. S.M. Kelly. Cresskill: Hampton Press.
Parsons, T. 1951. *The social system*. New York: Free Press.
Poli, R. 2009. The complexity of anticipation. *Balkan Journal of Philosophy* 1(1): 20–29.
———. 2013. A note on the difference between complicated and complex social systems. *Cadmus* 2(1): 142–147.
Prigogine, I., and I. Stengers. 1984. *Order out of chaos: Man's new dialogue with nature*. New York: Bantam Books.
Rasch, W. 1991. Theories of complexity, complexities of theory: Habermas, Luhmann, and the study of social systems. *German Studies Review* 14(1): 65–83.
Rasch, W., and E.M. Knodt. 1994. Systems theory and the system of theory. *New German Critique* 61(Winter): 3–7.
Rosen, R. 1985. *Anticipatory systems: Philosophical, mathematical and methodological foundations*. Oxford: Pergamon Press.
Shannon, C.E. 1948. A mathematical theory of communication. *The Bell System Technical Journal* 27(3): 379–423.
Taylor, M.C. 2001. *The moment of complexity: Emerging network culture*. Chicago: University of Chicago Press.

Teran, O. 2001. Review of 'How nature works: The science of self-organised criticality'. *Journal of Artificial Societies and Social Simulation* 4(4): 186–199.

Varela, F.J., and P. Bourgine, eds. 1992. Towards a practice of autonomous systems. *Proceedings of the first European conference on artificial life*. Cambridge: MIT Press.

Von Bertalanffy, L. 1972. The history and status of general systems theory. *The Academy of Management Journal* 15(14): 407–426.

Von Foerster, H. 1984. *Observing systems*, 2nd ed. California: Intersystems Publications.

———. 1990. Ethics and second-order cybernetics. Opening address for the international conference, *Systems and Family Therapy: Ethics, Epistemology, New Methods*. Paris, 4 October 1990. http://web.stanford.edu/group/SHR/4-2/text/foerster.html. Accessed 16 March 2015.

Von Foerster, H., and B. Poerksen. 2002. *Understanding systems: Conversations on epistemology and ethics*. New York: Kluwer Academic/Plenum Publishers.

Von Neumann, J. 1966. *Theory of self-reproducing automata*, ed. A.W. Burks. Urbana: University of Illinois Press.

Weaver, W. 1948. Science and complexity. *American Scientist* 36: 536–544.

Chapter 3
The Economy of Complexity

The full consciousness of the uncertain, the fortuitous, the tragic in all things human is far from having led me to despair. On the contrary, it is tonic to swop mental security for risk, since we gain opportunity thus... It is tonic to tear oneself away forever from the master word which explains everything, from the litany which pretends to resolve everything. It is tonic finally to consider the world, life, man, knowledge, action as open systems.

- *Edgar Morin (1973), Le Paradigme perdu: la nature humaine*

Abstract
An ontological understanding of complexity amounts to the view that complexity is irreducible for the reasons that there is no central organising principle, and complex systems are open. In this chapter, the notion of a complex, open system is reinscribed, and further explored, in terms of a theoretical understanding of economy.

The notion of economy, which denotes any constrained set of relations, was introduced into the philosophical literature by Bataille, who, in his work, distinguishes between a utilitarian or restricted economy and an excessive or general economy. In complexity terms, Bataille's central insight is the following: complexity is generated by a constrained set of relations, which gives rise to a system (the restricted economy of codified knowledge), but also to excess. The excess forms part of the system's environment (the general economy).

The exact manner in which the relation between the restricted and the general economies, or the system and its environment, should be conceptualised forms the grounds for different conceptions of economy. In this chapter, five views on economy are forwarded, namely: Hegel's totalising economy, Nietzsche's and Bataille's dual economy, Derrida's aporetic economy, and Nancy's immanent economy. The implications that these positions hold for understanding both systemic openness and relationality are explored at the hand of critical analyses in order to provide a philosophical account of the ontological view of complexity.

M. Woermann, *Bridging Complexity and Post-Structuralism*,
DOI 10.1007/978-3-319-39047-5_3

3.1 Introduction

In their article titled 'Towards an economy of complexity: Derrida, Morin and Bataille', Oliver Human and Paul Cilliers (2013) distinguish between two notions of economy: narrowly defined, an economy refers to 'the concern with the production and consumption of resources made possible by the orderly interplay of the parts of the system' (27). In this definition, it is assumed that resources are scarce, which—in turn—implies a constraint on the amount and type of production. In the broad, philosophical sense, an economy refers to *any* constrained set of relationships between the parts of a system; or, otherwise expressed, an economy is defined as a system that is '*dependent* upon the limits or constraints determined by the relationships between the components in the system' (27).

Following this broad definition, Human and Cilliers argue that—unlike *frame*—the notion of *economy* is, per definition, relational as opposed to atomistic (i.e. the unit of analysis is the *relation* between parts as opposed to the parts or the system as such). Furthermore, economies are necessarily studied as open systems because they stand in a constitutive relationship with their environments. This understanding of economy serves to challenge the inside/outside distinction that is conjured up by the term *frame*, and—to a lesser extent—by the term *system*. Given these two advantages of the notion of economy—i.e. the relational and open nature of economies—Human and Cilliers argue that "economy' is a useful placeholder to describe the 'slippery' nature of dealing with complexity' (39).

In this chapter, the nature of economy will be explored at the hand of Bataille's (1984; 1991) distinction between a restricted and a general economy, as introduced in the 'The notion of expenditure' and developed in *The Accursed Share*. It will be shown that Bataille manages to draw the problematic of economy into the human domain, by focusing on how organised relationality gives rise to both codified knowledge and to complexity. Briefly stated: for Bataille, a restricted economy is a utilitarian economy in which meaning can be defined and controlled, whereas a general economy lies beyond the domain of conceptual meaning, and therefore beyond our understanding and control. A restricted economy therefore deals with a knowable order (which is, at best, complicated in nature), whereas a general economy deals with an unknowable order, defined by excess and complexity.

Bataille has had a large influence on the post-structural discourse, and one of the reasons for his influence is that he introduces the terminology for dealing with open systems or economies. In so doing, his work raises important questions regarding the relation between the restricted and general aspects of any system or economy.[1] Bataille's understanding of economy was heavily influenced by his reading of

[1] See, for example, Derrida's 'From restricted to general economy: A Hegelian without reserve' (1978a) and Nancy's *The Inoperative Community* (1991), 'Concealed thinking' (2003a), and 'The unsacrificable' (2003b).

Nietzsche.[2] In the analysis that follows, Bataille's understanding of economy will be juxtaposed with the implied understanding of economy at work in Nietzsche's philosophy, and with the views on economy in both Derrida's and Nancy's philosophies. These thinkers will also be brought into conversation with one another in order to tease out the implications that the differences in their thinking hold for our understanding of complex systems, specifically relationality in complex systems.

The chapter will proceed as follows: In the first section, Bataille's notion of economy will be elucidated. In the second section, it will be shown that—although not explicated—the notion of economy also undergirds the Nietzschean project. In this regard, Nietzsche's influence on Bataille will be referred to. Specifically, it will be argued that—like Bataille—Nietzsche wishes to transcend the restricted economy; but, in so doing, his project culminates in an impossible ideal (i.e. the Nietzschean *Übermensch* or Overman). In the third section, Derrida's understanding of the nature of economy, and the double-logic that he employs to straddle the restricted and general aspects of economy, will be elaborated on with reference to his critical reading of Bataille, as well as his work on ethics and politics, law and justice, and violence and non-violence. It will be argued that the strength of his position lies in not privileging the restricted aspect of economy over the general aspect, but in showing how both are necessary to drive meaning forward. Nancy's views on community and thinking (as an expression of our finitude) will be introduced in the fourth section in order to demonstrate that the general economy is none other than the restricted economy. Throughout the chapter, the insights generated will be applied to our understanding of complexity with the explicit goal of putting forward a philosophical understanding of the economy of complexity.

3.2 Bataille

As mentioned above, Bataille's work on economy has had a large influence on the development of post-structural philosophy. In *Positions*, Derrida (1981: 106; 41) explicitly acknowledges this influence on his earlier works, in writing that these works are 'situated explicitly in relation to Bataille' to the extent that he works with 'a 'general economy,' a kind of *general strategy of deconstruction'*. In *The Inoperative Community,* Nancy (1991) engages with Bataille's work in an in-depth and sustained manner, especially with his thoughts on economy and community. Yet, although both philosophers acknowledge the space that Bataille accords to the general economy, they also question his conceptualisation of this space. More specifically, both Derrida and Nancy question the relation that Bataille poses between the restricted and the general economy or between codified meaning and complexity.

[2] In this regard, see the journal *Acéphale*, founded by Bataille and Pierre Klossowski; as well as Bataille's *On Nietzsche* ([1945] 1992).

3.2.1 The Restricted and the General Economy

In the preface to the *Accursed Share*, Bataille (1991) writes that the subject of his three volumes concerns 'a "general economy" in which the "expenditure" (the "consumption") of wealth, rather than production, was the primary object' (9). Unlike traditional economics in which a utilitarian analysis 'carried out to a view of a limited end, that of economic man' (23) is employed, Bataille's economics is concerned with the problematic of '*living matter in general*' in which 'energy is always in excess' (20). In this vein, he writes that '[o]n the whole a society always produces more than is necessary for its survival; it has a surplus at its disposal. It is precisely the use it makes of this surplus that determines it' (106). Bataille's work therefore signifies a shift from the perspective of the restricted economy to that of the general economy, which he maintains 'accomplishes a Copernican transformation: a reversal in thinking—and of ethics' (25).

Jürgen Habermas and Frederick Lawrence (1984) note that Bataille understands modernity in relation to a success-orientated utilitarian calculus, aimed at realising subjective ends. Bataille positions himself against this view by offering a moral critique centred on the unbounding of subjectivity in order to facilitate the return to the outlawed drives of spontaneous action, which, he believes, marks a subject's true liberation. Whilst Bataille's work is characterised by 'a base anti-idealist, non-teleological materialism' (Weiss 1986: 128), he nevertheless also seeks 'to link *material economics* (the production of material goods) and *theoretical economics* (the production of knowledge)' (Cutler Shershow 2001: 472). For this reason, and as noted by Jean Baudrillard (1987: 57), 'Bataille's thought goes, beyond proper political economy (which in essence is regulated through exchange value), straight to the metaphysical principle of economy'. It is this theoretical order, rather than a narrowly-defined economic or political order, that is of interest in this analysis—especially to the extent that Bataille's economics raises questions of meaning and knowing in complex systems.

Derrida understands Bataille as suggesting that all traditional philosophical systems of thought should be considered as restricted theoretical economies (Shershow 2001). Like traditional economic systems, thought systems are necessarily limited by the frame of analysis that is employed, which ultimately leads to the generation of knowledge from within a limited perspective. Poli (2013: 146) argues that '[s]cience is for the most part a set of techniques for closing open systems in order to scrutinise them'. This same point applies to knowledge generation in general, with the Luhmannian consequence that a strict demarcation is drawn between the knowable and controllable inside of the system, and the unknowable outside. All systemic excesses are relegated to this outside, and the non-utilitarian forces that act upon the system are not accounted for by the system itself. In contrast to this traditional view of economy, Bataille's interest concerns unproductive expenditure, in which meaning generates a surplus of signification that cannot be controlled or mastered, and that Bataille (1991: 10) moreover translates as 'the effervescence of life'.

To this end, Bataille positions himself in relation to the Hegelian project, as articulated in *Phenomenology of Spirit*, and offers a critique of Hegel's (1977) totalising idealism. In Hegel's system the iterative dialectical operation between thesis and antithesis ultimately gives rise to a grand synthesis or *Aufhebung*. Bataille is specifically concerned with the problems of 'the reduction of the other to the same, the all-encompassing nature of philosophical reason, and the end of philosophy' (Trahair 2001: 156) that characterises Hegel's work. Bataille poses a distinction between knowing (*savoir*) and un-knowing (*non-savoir*), or reason and unreason. Whereas he identifies Hegel with the former, his own philosophical project is concerned with the latter. Otherwise stated, and as noted by Lisa Trahair (2001: 162), 'Bataille therefore contrasts between the restricted economy that characterises the circulation of meaning in the *Phenomenology* and the general economy that exposes meaning to its comic underside, that wastes meaning, destroys it without reserve.'

However, Trahair (159) also argues that Bataille's 'philosophy of un-knowing [which destroys meaning] is in no way a celebration of ignorance. It is, rather, a response, a very precise interjection in relation to Hegel's thought'. This interjection is premised on Bataille's understanding of experience. Whereas for Hegel, experience is dialectically related to self-revelation; for Bataille, experience is a non-teleological and sustained experience of the unknowing (laughter, drunkenness, erotic and sacrificial effusion, poetic and heroic behaviour, anger, and absurdity all serve as examples on the continuum of un-knowing). Trahair furthermore draws on Nick Land's (1992) and Joseph Libertson's (1982) commentaries on Bataille in order to demonstrate how the experience of un-knowing challenges the formal Kantian and speculative Hegelian understanding of the noumenon.

For Kant (1929), phenomena are appearances in the world that we know through our sensory organs; whereas noumena, which are inaccessible to the senses, are the things-in-themselves (*das Ding an sich*). Understood positively, noumena allow us to experience phenomena, even though they remain conceptually empty.[3] The noumenon is therefore viewed as the basis for thinking from which phenomena emerge. Land however argues that Bataille's 'fanged noumenon' represents not the beginning, but the end, of knowledge. The un-knowing constitutes a 'slide into oblivion' or a 'dissolvent immanence' (Land 1992: 116). Since we cannot know things-in-themselves our phenomenological condition is one of incomprehension because our grasp on the world is tenuous. This incomprehension can be expressed in terms of heteronomy or heterogeneity; which, for Bataille, refers to that which is nonsensical from the limited perspective of a restricted economy, and which cannot be included in this economy without destroying its radical status[4] (Human and

[3] Maturana's work on the perceptual system and colour vision of frogs cited in the previous chapter also validate this point from a scientific perspective.

[4] Habermas and Lawrence (1984: 85) note that, for Bataille, 'the heterogeneous is related to the profane world as what is superfluous—from refuse and excrement, through dreams, erotic temptations and perversions, to contaminating, subversive ideas; from palpable luxury to exuberantly electrifying hopes and transcendences. In contrast to this, the homogenous and conformist elements of everyday life are the result of profane processes of production and exchange.'

Cilliers 2013). Many anti-intellectualist thinkers (including Nietzsche) have grappled with heterogeneity, which denotes not only the inability of a restricted economy to represent alterity, but also the manner in which alterity alters thought (which is no longer seen as operating solely in service of meaning).

In order to better understand Bataille's view of alterity or negativity, as well as the manner in which it threatens Hegel's entire teleological project in which meaning and experience are conserved, it is necessary to briefly investigate the difference between Hegel's notion of lordship and Bataille's notion of sovereignty.

3.2.2 Hegel Versus Bataille, or Lordship Versus Sovereignty

The differences between Hegel and Bataille can also be described in terms of the differences between lordship and sovereignty. The notion of lordship is treated in Hegel's *Phenomenology of Spirit* at the hand of the dialectic between master and slave, which plays a prominent role in the development towards self-consciousness. The famous Hegelian scholar, Alexandre Kojève (1969: 3), writes that man becomes conscious of himself 'at the moment when—for the "first" time—he says "I"' (in other words, when he becomes conscious of his human reality and dignity). However, each 'self-consciousness' can only be certain of itself once another human recognises it as authoritative. The dissolution of the isolated 'I' is therefore the result of the first experience or the first '(murderous) fight' (15) with another, in which both parties try to force the other to recognise him/her as a self-conscious subject, whilst withholding reciprocal recognition. The struggle ends when one party triumphs as an independent consciousness existing for itself (Kohn 2005). The other party chooses submission and life over death, thereby accepting its essential reality as animal-life, i.e. given-being for another entity. Whereas the former party is the master or lord, the latter party is the slave or bondsman (Kohn 2005; Kojève 1969). Hegel (1977: 115) describes the implications of this struggle as follows:

> The lord is the consciousness that exists *for itself*, but no longer merely the Notion of such a consciousness. Rather, it is a consciousness existing *for itself* which is mediated with itself through another consciousness, i.e. through a consciousness whose nature it is to be bound up with as an existence that is independent, or thinghood in general. The lord puts himself into relation with both these moments, to a *thing* as such, the object of desire, and to consciousness for which thinghood is the essential characteristic.

The outcome of the struggle creates the dialectic between master and slave. On the one hand, the slave (in wanting to conserve his life) is held in the master's domination, thereby providing the master with self-conscious certainty. This is because the master's reality is no longer subjective and immediate, but objectified and mediated by the slave's recognition. The master however recognises the slave's reality as a consciousness for which 'thingness' is the essential entity (in that the slave, who refuses to risk, binds himself completely to the things on which he depends). The master is related in a mediated fashion to the 'thing as such' or the object of desire, through means of the slave. This is because it is the slave's job to transform natural objects or raw materials, with the view of their consumption by the master (Kojève 1969).

Derrida (1978a: 254) defines lordship as 'an obligatory stage in the history of self-consciousness and phenomenality' and thus 'in the presentation of meaning'. However, lordship nevertheless remains trapped within the dialectic, since the lord or master needs the bondsman or slave to confirm his independent self-conscious existence: 'The *truth* of the independent consciousness is accordingly the servile consciousness of the bondsman' (Hegel 1977: 117). This is because lordship is premised not only on life, but also on the slave. The truth of the lord is the slave, and the master must do work to keep the slave's consciousness repressed in order to experience pure self-consciousness. Therefore, the master's consciousness 'is not an autonomous Consciousness, but all to the contrary, a dependent consciousness, that exists for [the slave]' (Kojève 1969: 20).

The tragedy of the master is that the master's being *for itself* is premised on the slave's recognition of his essential being. However, the slave (whose consciousness remains 'thingness') is not a being worthy of respect and dignity. Thus, as Kojève (19) explains:

> The master has fought and risked his life for a recognition without value for him. For he can be satisfied only by recognition from one whom he recognizes as worthy of recognizing him. The Master's attitude, therefore, is an existential impasse. On the one hand, the Master is Master, only because his Desire was directed not toward a thing, but toward another desire—thus, it was a desire for recognition. On the other, when he has consequently become Master, it is as Master that he must desire to be recognized; and he can he recognized as such only by making the Other his Slave. But the Slave is for him an animal or a thing. He is, therefore, "recognized" by a thing. Thus, finally, his Desire is directed toward a thing, and not—as it seemed at first toward a (human) Desire.

Ironically then, the master is enslaved by the consciousness of the bondsman, who represents an animal or thing, as opposed to the independent consciousness of an equal. Servility is therefore the condition of lordship (Derrida 1978a). Unlike the master however, the slave can transcend lordship. Kojève (1969: 20) explains that the slave 'as *repressed* Consciousness' can 'go within itself and reverse and transform itself into true autonomy'. The way in which the slave can overcome his slavery is through work, and by establishing a relationship with work which, as Derrida (1978a) and Kojève (1969) explain, is characterised by the deferral of pleasure. The master immediately negates desire and, hence, life in pleasurable consumption. The slave, in contrast, must inhibit his desires (work is repressed desire) and delay the disappearance of the thing (by first transforming it through work). Through transforming things in the world, the slave is thus able to transform himself into a formed and educated man. Therefore, in summary:

> just as lordship showed that its essential nature is the reverse of what it wants to be, so too servitude in its consummation will really turn into the opposite of what it immediately is; as a consciousness forced back into itself, it will withdraw into itself and be transformed into a truly independent consciousness (25).

Although the dialectic between master and slave is radically revised through the course of Hegel's analysis, it nevertheless remains intact. This is because work and the repression of desire constitute a form of sublative negation, whereby man 'tarr[ies] with the negative' (Hegel 1977: 19). Derrida (1978a: 254) describes sublative negation as 'the negation characteristic of consciousness, which cancels in such a way that it preserves and maintains what is sublated'. By putting negativity to work in service of self-consciousness and of meaning, and by defining it as a necessary link in the chain of phenomenology, Hegel is able to develop a productive conception of negativity that still operates in service of the dialectic. Derrida (255–256) concludes that:

> Through this recourse to the *Aufhebung,* which conserves the stakes, remains in control of the play, limiting it and elaborating it by giving it form and meaning (*Die Arbeit... bildet*), this economy of life restricts itself to conservation, to circulation and self-reproduction of meaning.

It is specifically this economy that Bataille wishes to disrupt. Bataille's philosophy constitutes an attempt to break out of the totalising Hegelian system. This is achieved through recourse to a third term, namely that of sovereignty. Sovereignty finds its expression in abstract negativity, i.e. 'the absolute loss of meaning' or 'mute and non-productive death' (255). Derrida maintains that the central contribution of Bataille's philosophy concerns this re-conceptualisation of negativity. It is precisely because of this conception of abstract negativity that Bataille can pull sovereignty out of the dialectic in order to free it from the stakes and cause a disruption that prevents the system from playing out in terms of the Hegelian *Aufhebung*. Sovereign transgression (as an 'experience' of abstract negativity) is the singular event in which meaning sinks and vanishes, before being taken up again in the dialectic. In this regard, Derrida (256) writes that Bataille's negativity 'never takes place... never *presents* itself, because in doing so it would start to work again.'

Wherein does meaning sink and vanish? According to Derrida, it disappears in a burst of laughter that never literally appears because it exceeds the possibility of meaning. This burst of laughter reverberates on the continuum of un-knowing, which can further be invoked to explain the difference between Hegel's and Bataille's positions. In this regard, Derrida (263) writes:

> The *continuum* is the privileged experience of a sovereign operation transgressing the limit of discursive difference... Pushing itself towards the nonbasis of negativity and of expenditure, the experience of the *continuum* is also the experience of absolute difference, of a difference which would no longer be the one that Hegel had conceived more profoundly than anyone else: the difference in the service of presence, at work for the history (of meaning). The difference between Hegel and Bataille is the difference between these two differences.

Even though sovereignty cannot be spoken (it is the burst of laughter that never literally appears), Bataille also insists that it must be spoken. To make sense of this statement, it is useful to follow Carolyn Bailey Gill (1997) in her exposition of the 'Summit and decline' section of Bataille's (1992) *On Nietzsche*. Herein, Bataille juxtaposes two summits, namely the *immediate summit* and the *spiritual summit*.

Bataille (17) writes that the immediate summit 'corresponds to excess, to an exuberance of forces… It relates to a meaningless expenditure of energy and is a violation of the integrity of individual beings.' It is defined as 'happiness at the brink of the abyss', it ruptures ordinary time, and it denotes a being's glorious loss.[5] The immediate summit, in other words, is a transgression of the restricted economy; and, in this regard, Bataille (1992: 33) writes that 'this blessed openness isn't humanly imaginable'.[6]

However, this unendurable and immediate summit always (and simultaneously) gives way—or slides—into its double, namely the spiritual summit, which is defined by a loss of strength, whereby 'we condemn excesses in the name of some higher ideal… [we become] preoccupied with gaining goods of all kinds, acquiring wealth, since we're thinking about the difficulties to come' (39). Bailey Gill (1997: 94) notes that the spiritual summit is related to limits and prohibitions that stem from our finitude. Sovereignty which cannot be spoken, but which must be spoken, is therefore always invoked when it is already in decline, i.e. any attempt to explain the immediate summit (the experience of the general economy, and therefore of chance and impossibility) is always only from the point of view of the spiritual summit (i.e. the restricted economy of necessity and possibility). Sovereignty is thus passive in that it lies outside of dialectics. Yet, it nevertheless imposes itself on reason; it is felt by reason.

The relation between the restricted economy and the general economy can also be expressed in terms of the difference between lordship and sovereignty, the immediate summit and the spiritual summit, meaning and non-meaning, or codified knowledge and complexity. Although the sovereign operation has no meaning, it is nevertheless not meaningless. In this regard, Derrida (1978a: 256) argues that the importance of the sovereign is that it makes 'the seriousness of meaning appear as an abstraction inscribed in play'.

3.2.3 The Possibility of Open Economies

At this juncture, it is necessary to flag some of the difficulties resulting from Bataille's understanding of the restricted and the general economies, especially since these difficulties also characterise any serious engagement with complexity. In the closing paragraphs of her article, Bailey Gill (1997: 96) notes that: '[p]erhaps Bataille no more than Nietzsche (no more than Derrida) can escape the serious difficulty of thinking the beyond of a horizon of expectation, or … of thinking an indifference to the contents of the future'. Although she mentions this only in passing,

---

[5] In Derrida's (1978a: 265) words: 'sovereignty has no identity, is not *self, for itself, toward itself, near itself*… [I]t must… lose itself, lose consciousness, lose all memory of itself and all interiority of itself.'

[6] Bailey Gill (1997) interprets this passage as a reference to Nietzsche's *Übermensch*, which is treated in Sect. 3.3.1.

and makes no attempt to address the issue raised, I would argue that this remark captures the central difficulty of the Bataillian project: namely, how to think productively about the general economy; or, in complexity terms, how to conceptualise systemic openness.

Scott Shershow (2001: 489) frames this difficulty as follows: '[a]ttempts to relate the restricted and general economies seemingly always return to this dilemma: the necessary restrictedness involved in ... economics itself ... apparently also requires, as it ground and vehicle, a restricted economy of knowledge'. In other words, if we follow Human and Cilliers (2013) in their understanding of an economy as necessarily *dependent* on limits and constraints, then Bataille's general economy cannot be conceptualised or understood as the other of the restricted economy. Although Bataille attempts to structure his analysis around waste and excess, these terms become the very restriction that delimits his philosophical project. Bennington (1995: 47–48) succinctly sums up the implications of this, in writing that:

> In its most abstract form, this suggestion would say that 'general economy' is not the other of 'restricted economy,' *but is no other than* restricted economy; that there is no general economy except as the economy of the restricted economy; that general economy is the economy of its own restriction.

According to this logic, any attempt to conceptualise an open system (or to relate a system to its environment) would necessarily be restricted by the concepts employed; it would be intelligible as an economy of meaning, and would therefore denote an informationally-closed system, albeit one that employs a broader unit of analysis. Niklas Luhmann (2000) also remarks on this problem in arguing that drawing a distinction between the inside and the outside of a system is a necessary condition for observing and describing identities. He argues that these distinctions form a systemic frame, which he describes as 'the self-produced and reproduced difference of the system and its environment' (46). This frame, in turn, produces informational closure, since all distinctions are necessarily made from the internal perspective of the system: 'the system can make distinctions and thereby frame its own (but only its own!) observations' (46). In order to describe the system as such, we therefore need a *Theory of Frames*. This however creates a self-referential paradox, which can be expressed as follows: in trying to distinguish clearly between the system and its outside, we reproduce the distinction between inside and outside within the system itself, thereby creating information distortions. Luhmann's point is that there cannot be a meta-frame (in post-structural terminology, a theory of economy) that can objectively describe both the identity of the system as such, and the relation between the system and its environment.

Bataille was certainly aware of this problem in describing sovereignty as that which cannot be said. However, he also recognised that in order to be *meaningful*, we need to be able to *say* something about the general economy. In the introduction of this chapter, I referenced Human and Cilliers' (2013) argument that the terminology of *economy* is preferable to that of *frame* because economies challenge the inside/outside distinction and necessarily denote open systems. Whether this is indeed the case—i.e. whether one can think differently than Bataille and Luhmann about the relationship between a restricted and a general economy, or between the inside and the outside of the system—will be the central problem that will be addressed in the remainder of the analysis in this chapter.

Essentially, this problem amounts to a different version of the question raised in the previous chapter, namely whether there is a productive way in which to reconcile our understanding of informationally or operationally-closed systems and open systems. In response to this question, Morin's elucidation of why the principle of organisation necessitates both systemic openness in operationally-closed systems and operational-closure in open systems was referenced. However, in this analysis, the focus shifts from the physical domain to the phenomenological domain, with the consequence that systemic openness and closure take on different definitions. Essentially, systemic openness refers to a system that is capable of accommodating alterity, without assimilating this alterity into the logic of the system; whereas systemic closure denotes a system in which alterity cannot be accommodated within the system, except as an observation generated by the system itself.

The analysis will commence with an exploration of the implied use of economy in Nietzsche's work, for the reason that Nietzsche has had a profound influence on Bataille's work in particular, and the development of Western philosophy in general. In this regard, Foucault (1974: 37–38) notes that Nietzsche dedicated us to 'a form of thought… which would be, absolutely and in the same motion, a Critique and an Ontology, an understanding that comprehends both finitude and being'.

3.3 Nietzsche

Trahair (2001) notes that, until Nietzsche arrived on the scene, philosophy and reason were happily celebrating the honeymoon period of their marriage. Alain Badiou (2001: 1; 3) writes that 'Nietzsche is not a philosopher, he is an anti-philosopher… To enter into Nietzsche one must… focus on the point where evaluation, values, and sense all come to falter in the trial posed by the act'. Badiou describes this act (which is without precedent) as something 'that will in fact destroy philosophy' (1) and he further states that the trial posed by the act characterises that which 'is no longer a question of values or of sense, but of what actively surpasses them, what philosophy has always named "truth"' (3). Nietzsche's challenge to the happy union between philosophy and reason, and his anti-philosophical stance towards all our philosophical categories is 'felt' in his views on superhuman self-assertion and fatalism, as well as in his questioning of human aims and his celebration of the will to power. These themes also reveal Nietzsche's preoccupation with the general economy of meaning.

In this section, I shall therefore briefly explore these ideas, and also attempt to position them in relation to Bataille's reading of Nietzsche. The analysis will begin with an overview of Nietzsche's *Mensch* (human) and *Übermensch* (Overhuman), followed by a discussion on the *Eternal Return*, and will end with an economic reading of the Nietzschean project.

3.3.1 *Nietzsche's* Mensch *and* Übermensch

Nietzsche's view of the human is, as noted by Vasti Roodt (2005: 84), based on a response to his view of modernity as 'the loss of the human' and the transvaluation of all values. This view is expressed in one of the most famous and misunderstood passages in *The Gay Science,* in which Nietzsche (1974: 181) proclaims that 'God is dead. God remains dead. And we have killed him.' This passage refers to the 'crisis of the human being' (Gillespie 1999: 144) that characterises the modern view of nihilism, and that is tied to both 'the degeneration of God and the degeneration of the human' (Roodt 2005: 84), as a fixed measure of the world and as a product sculpted in the image of a God.

Rather than to simply accept this predicament of modern nihilism, Nietzsche attempts to rethink the nature of man in the absence of a transcendental authority. He does this by framing identity in terms of a process of becoming, and by conceptualising meaning in terms of multiplicity, indeterminacy, or excess. Roodt (87) notes that Nietzsche is able to avoid the reification of human becoming into being by adopting 'the crab-like strategy of the genealogist', which means tracing 'the network of shifting relations that tell the story of how we have become—and are becoming—who we are.' She further argues that, for Nietzsche, the human is 'a temporary constellation of forces under the rubric of the will to power' (87) that manifests in this world of appearances that constitute our reality. Reality therefore is nothing other than the sum total of a dynamic interplay of forces or the fight between power-complexes. In another famous pronouncement from *Beyond Good and Evil,* Nietzsche (1989a: 48) describes the ontology of the world as follows: 'The world viewed from inside, the world defined and determined according to its "intelligible character"—it would be "will to power" and nothing else.—'

Despite his view of humans as a 'particular constellation of forces within the force-field that is the will to power' (Roodt 2005: 102), Roodt also notes that, for Nietzsche, 'a mere collection of drives or power-quanta does not yet constitute a human being' (105). In addition, what is needed is some type of organising principle that can order the various power complexes of the human, without destroying the tensions and contradictions between these power complexes. Paul van Tongeren (2000: 236) describes this ordering as the impossible ideal of living out the 'contradiction between engagement and disengagement, between partiality and... supra-partiality.' This ordering, which Nietzsche views as an exercise in imparting style, is also what is necessary for us to become who we are (see Nietzsche 1974: 263–266). In this regard, he remarks: 'The wisest man would be the one richest in contradictions, who has, as it were, antennae for all types of men—as well as his great moments of *grand harmony*—a rare accident even in us!' (Nietzsche 1967: 150).

For Nietzsche, becoming who we are is also a test of endurance, through which we strive to reach an impossible ideal. The injunction to become one's self therefore marks an absolute limit, which—to varying degrees—one approximates. Whereas the lowest human would try to repress or reconcile the contradictory forces in the interest of attaining a perfect self-understanding, van Tongeren (2000: 242) writes that '[t]he highest human would realize the greatest multiplicity in the most tension-fraught but "controlled" manner.' The absolute limit, denoted by the Nietzschean concept of the *Übermensch* however remains other-worldly. Both Roodt and van Tongeren argue that the Nietzschean ideal is non-teleological to the extent that it does not present the highest stage of humanity, but rather something which 'exceed[s] the limits of humanness, [that is] different from "menschlich"… a being that is beyond the human being' (244; 245).

Support for this interpretation can be found in the following passage, wherein Nietzsche writes that '[i]t is *absolutely* not the aim, that the latter ones [the *Übermenschen*] be conceived of as the masters of the first ones [the last men]: but: there should be two types next to each other—as much as possible separated' (Nietzsche 1999a: 244; trans. van Tongeren). This description begs the question as to what role the figure of the *Übermensch* as the other-to-man plays in the philosophy of Nietzsche. In answer to this question, van Tongeren (2000) argues that the *Übermensch* is one of Nietzsche's artistic creations aimed at, on the one hand, expressing his ideal of a complete affirmation of conflicting power complexes; and, on the other hand, of attempting to overcome the seeming impossibility of its full manifestation. Roodt (2005), in turn, postulates that the *Übermensch* is, firstly, Nietzsche's reaction to the epithet that modern man (and his self-preservation) represents the terminus of world history, with the consequence that man becomes the measure of all things. Secondly, she argues that the '*Übermensch* is a battle-cry against the ascetic ideal' (116), in that the meaningless of the world experienced as nothing more than the fact of existence is replaced by a higher ideal. In the words of Nietzsche: 'Not "mankind" but *overman* is the goal!' (Nietzsche 1967: 518).

The notion of the *Übermensch* is closely related to Nietzsche's view of the *Eternal Return*, which should not be interpreted as a cosmology, but as an affirmation of his ontology. In brief, and at the hand of Roodt's exposition of the subject, the Eternal Return represents a response to the notion that the world and meaning operate in service of an external principle or a higher reality (whether conceived of as an ideal, creator, or truth). Nietzsche attempts to overcome this teleological order in which our lives, and the world in general, are conceived of in terms of a utilitarian calculus ('life/the world for the sake of x'). He abandons this moral interpretation (this higher measure) of comparison and judgement, in favour of the unconditional affirmation of everything that exists, and therefore in favour of the world itself. Overcoming this moral interpretation requires that we acknowledge that 'nothing is self-sufficient, neither in us ourselves nor in things' (532). Everything, in other words, is relationally bound to everything else in a force-field of power-wills, which makes it impossible to deny even one moment of existence. In this regard, Nietzsche (1969: 331–332) writes:

> Did you ever say Yes to one joy? O my friends, then you said Yes to *all* woe as well. All things are chained and entwined together, all things are in love; if you ever wanted one moment twice, if ever you said: 'You please me, happiness, instant, moment!' Then you wanted *everything* to return! you wanted everything anew, everything eternal, everything chained, entwined together, everything in love, O that is how you *loved* the world…

The Eternal Return is shorthand for the idea that every instance is made possible by what has gone before, and is the condition for what is still to come. In Roodt's (2005: 197) words: '[t]he 'moment'... becomes the knot to which everything is tied, so that all that has been and all that will be resounds again and again in every instant.' As such, the ultimate purpose of the world does not reside in an order beyond the world, but in every moment. We are therefore irrevocably tied to our fate, and the highest form of affirmation is love of this fate:

> My formula for greatness in a human being is *amor fati*: that one wants nothing to be different, not forward, not backward, not in all eternity. Not merely bear what is necessary, still less conceal it—all idealism is mendaciousness in the face of what is necessary—but *love* it (Nietzsche 1989b: 258).

What is therefore required is not merely to accept, but to *will* the world as it is, in all its irreducible plurality. This capability—which fate has either granted or denied each of us to a lesser or a greater extent—requires that we follow the path of nihilism to its very end, and accept a vision of the world without purpose. It requires of us to 'think this thought in its most terrible form: existence as it is, without meaning or aim, yet recurring inevitably without any finale of nothingess: "*the eternal recurrence*."' (Nietzsche 1967: 35). Although this vision demands that we forsake all value, it also offers redemption in returning us to the world, and is therefore the greatest act of affirmation. However, and as noted by Roodt (2005), the dissolution of distinctions between the instant and the eternal, what is and what ought to be, and the particular and the universal, also present us with an unrealistic vision of redemption in that it is not fully attainable by mere mortals. Nietzsche therefore links this redemption to the beyond of the human, which in one characterisation, is signified by the figure of the *Übermensch*, whose affirmative vision we can approximate but never embody. As Roodt (201) remarks: 'Nietzsche intensifies and radicalises the new beginning [made possible by the full affirmation of the Eternal Return] to such an extent that it leaves the human behind.'

3.3.2 Nietzsche's Economy

Returning to Bataille's distinction between a restricted and general economy, it should be clear at this juncture that Nietzsche's project constitutes a critical response to the restricted economy that marks modern nihilism. As a result of the denunciation of a higher order or external ideal, the closed economy of modernism collapsed under the force of its own pointlessness. In other words, if everything is reducible to a knowable and controllable closed order, then the system invariably plays out, culminating in the figure of the last man, who is the achieved goal of world history. The same point can be made from the perspective of GST, in which it is recognised that complex, living systems function under conditions of nonequilibrium or energetic flux. In other words, no capacity for complex behaviour exists within a fully constrained system, which means that a fully constrained system will result in system's death. In order to avoid this nihilistic outlook in which meaning is fully determinable and, hence, ultimately meaningless; Nietzsche reinscribes the human and the world in terms of a force field of contradictory and opaque power relations, by which the human becomes non-transparent to herself. Paradoxically, it is in the murkiness of our own identities and the identity of the world wherein we can find meaning.

Nietzsche therefore reintroduces complexity into ontology, or otherwise stated he creates the possibility for the excesses of the general economy to infiltrate the restricted economy. Our ability to accommodate the excesses depends on our ability to approximate the ideal limit, which, as previously argued, is constituted by the Eternal Return of a great multiplicity that does not collapse under the force of its own contradictions. For Nietzsche, it is the figure of the philosophical prophet—exemplified in his fictional character Zarathustra—who comes closest to accepting this ideal. And, yet, Nietzsche still draws a definitive line between man and the one who exists beyond man, as is clear from the following passage in *Zarathustra*:

> The most cautious people ask today: 'How may man still be preserved?' Zarathustra, however, asks as the sole and first one to do so: 'How shall man be *overcome*?' The Superman lies close to my heart, *he* is my paramount and sole concern – and *not* man: not the nearest, not the poorest, not the most suffering, not the best (Nietzsche, 1969: 297).

It is interesting to note that although Bataille affirms the idea of the Eternal Return, he does not wholly agree with Nietzsche's interpretation thereof. Bataille criticises Nietzsche for his focus on will to power, as he does not believe that this idea can do justice to sovereignty. For Bataille, the death of God and the loss of a transcendental point of reference should not be understood as man's dissolution in universal becoming, whereby 'the world concentrates itself in man' (Fink 2003: 213) as Nietzsche would have it, but as '"the fall of the universal into humanity" and the fall of humanity into sheer chance' (Weiss 1986: 132). In other words, Bataille shifts the focus from the will to power to the will to chance,[7] for the reason that he believes that true sovereignty is not an act of power but an act of rebellion that cannot be reduced to interiority, and hence individual will (138). In this regard, Bataille (1970: 23, trans. A.S. Weiss) writes in *Sur Nietzsche*:

> I believe that we must, in this sense, reverse the idea of the Eternal Return. Our anguish derives, not from the promise of infinite repetition, but from the following: the moments grasped within the immanence of return suddenly appear as ends. Remember that *in all systems* those instants are considered and assigned as means; morality always says: "let every instant of your life be *motivated.*" The Return *de-motivates* the instant, frees life from purpose and is thereby, first of all, its downfall. The Return is the whole of man's dramatic mode and his mask; it is the desert of a man whose every instant is henceforth unmotivated.

[7] Bataille (1992) states in *On Nietzsche*, which is partly written in diary form, that it is only with his life that he could write a book on Nietzsche, and that his book is, in part, an account of the daily casting of dice.

It is precisely the unmotivated nature of the instant that makes it foreign to will, and that allows the moment to stand in service of chance[8] and the sovereign. However, Allen Weiss (1986: 139) argues that framing the Eternal Return in terms of the dramatic mode reifies the instant and transforms it into a sign ('a mere simulacrum of subversive demotivation—which transfigures the subject, but nevertheless asserts it as dramatis persona'). Therefore, although Bataille wishes to describe the Eternal Return in terms of a heterogeneous and incommunicable experience of sheer chance, his description of chance in terms of a very precise interpretation of the Eternal Return does not allow for the disruption of memory, which Weiss deems as a necessary condition for giving oneself over to chance. This issue represents another version of the central problem articulated in Sect. 3.2.3, namely, whether it is possible to describe the workings of the general economy from within the perspective of a restricted economy.

As with Bataille, Nietzsche also struggles to successfully overcome this problem. Although Nietzsche affirms the workings of the general economy in the world (which is typified by a constellation of power struggles that generate excess and indeterminacy, and that point to the limitations of any utilitarian analysis), he poses the ideal of the general economy as something that cannot manifest in this world. Nietzsche is correct in thinking that this ideal cannot be realised in the world of man, since what Nietzsche requires of us is to engage with the incompressible nature of complexity, which is impossible given our limited vantage points and the inherent limitations of all meaning or signified content. The sober recognition of complexity in all its fullness requires the type of objective metaframe that is simply impossible due to the necessarily bounded (and hence differentiated and exclusive) nature of all systems of meaning. Furthermore, boundaries are necessary for identity formation, since although they delimit a particular characterisation of a system, they are also *constitutive* of a system (Cilliers 2001). Cilliers (141) therefore writes that boundaries are not only the product of the description that we give to the system under study, but are also 'a function of the activity of the system itself'. Identity is therefore contingent on boundary formation, which begs the question as to what type of non-creature Nietzsche's *Übermensch* might be?

Although Nietzsche's philosophy constitutes a powerful critique of transcendence in favour of a radical immanence, the fact that his *Übermensch* remains otherworldly means that, on the final count, he—like Bataille—still works with a dual ontology: one which is meaningful, limited, restricted, and therefore still an

[8] It should be noted that Nietzsche was not blind to the workings of chance. In a passage from *Daybreak,* Nietzsche (1982: 80; 81) writes: '*Purposes? Will?*—We have accustomed ourselves to believe in the existence of two realms, the realm of *purposes* and *will* and the realm of chance… [but] all is not purpose that is called purpose, and even less is all will that is called will! And if you want to conclude from this: 'so there is only one realm, that of chance accidents and stupidity?' – one will have to add: yes, perhaps there is only one realm, perhaps there exists neither will nor purpose, and we have only imagined them.'

example of economy; and one which is unrestricted, which leaves the human behind, and which we essentially cannot philosophise about. The fact that Nietzsche construes the *Übermensch* rather than the *Mensch* as the final goal points to the impotence of his philosophical vision, and does not allow for a productive engagement with the idea of the general economy. This point will become clearer on further exploration of Derrida's use of economy, as elaborated upon in the next section.

3.4 Derrida

3.4.1 The Aporetic *Nature of Economy*

Derrida affirms Bataille's distinction between a restricted, calculable view of meaning and an excessive, non-calculable view of meaning. However, what Derrida contests is Bataille's assignation of these two views on meaning to two separate economies—one restricted, and the other general. Rather, Derrida is of the view that excessive, incalculable meaning resides in our own contexts, and not in an order beyond the world. Indeed, there can be no pure general economy, if we—to re-cite Human and Cilliers (2013 27)—understand *economy* in the philosophical sense as something that 'is *dependent* upon limits or constraints determined by the relationships between the components in the system.' For Derrida, the only manner in which to think about the general economy is by considering the *aporia* or dilemma that exists between utilitarian and non-utilitarian forces. In this regard, he writes:

> This unnameable is not an ineffable Being which no name could approach: God, for example. This unnameable is the play which makes possible nominal effects, the relatively unitary and atomic structures that are called names, the chains of substitutions of names in which, for example, the nominal effect *différance* is itself enmeshed, carried off, reinscribed, just as a false entry or a false exit is still part of the game, a function of the system (Derrida 1982: 26–27).

What should become clear during the course of this analysis is that—unlike Nietzsche's and Bataille's engagements—Derrida's engagement with the problematic of economy does not hinge on the ideas of a meaningful 'inside' and a foreign 'outside' (what is denoted by 'an ineffable Being' in the above citation). For Derrida, systems are meaningful, but their meaning cannot be exhausted, due to the reason that systems of meaning are inherently open. In order to motivate this claim, I turn to Andrea Hurst's (2004) systematic unpacking of what she calls the *logical "form"* that is iterated throughout Derrida's texts. She firstly explains this logical form in terms of the meanings of *différance*, and then shows how these meanings are themselves constitutive of the *aporia* that marks Derrida's account of meaning.

Différance, she explains, must be thought of negatively, which however does not mean that it should be thought of in terms of absence. Rather: on the one hand, *différance* should be understood in terms of 'a regulated economy of difference and deferral' (254). It is 'the process of scission and division which would produce or constitute different things or differences' (Derrida 1982: 8–9); or, in complexity terms, it is the act of drawing distinctions in order to frame our observations (Luhmann 2000). On the other hand, *différance* is an aneconomic term, 'an expenditure without reserve, as the irreparable loss of presence... that apparently interrupts every economy' (Derrida 1982: 19). In complexity terms, this meaning of *différance* concerns the play of disorder, anti-organsiation, and entropy with the system. The relation between the economic (or restricted) and the aneconomic (or general) dimensions of *différance* is neither oppositional nor dialectical, but *aporetic*. This means that these terms are absolutely irreconcilable, but equally necessary for the generation of meaning. However, Derrida's analysis also suggests that these terms are related asymmetrically. The economic is unavoidable as long as there is life, but the aneconomic is always stronger. Therefore, an absolute collapse is possible (although without witness). Yet, ruin (and the possibility of total ruin) also, paradoxically, inaugurates the possibility of meaning (which is to say that ontology can always be deconstructed). In Morin's work on complexity, the order is reversed: to recall,[9] although order and organisation copulate with disorder and anti-organisation to create meaning, Morin (1992: 49; italicised in the original) argues that, once established, '[o]rder and organization... are capable of gaining ground on disorder.' And, yet, the potential for the radical transformation (and even the collapse) of complex systems also exists. This suggests that order and organisation are not an inevitable outcome. Meaning (and the future) always manifests in a contingent, as opposed to a necessary, manner.

In trying to think together the restricted and general dimensions of meaning, Derrida's logic aims to transgress the limitations that our traditional binary logical schema (which is necessary restricted) places on us, and for this reason his thinking can be described as complex. This implies that Derrida cannot simply be thought of an anti-foundationalist. Indeed, he emphatically supports a foundational logic to the extent that he deems it a necessary condition for conceptual language. In his words: 'Every concept that lays claim to an rigor whatsoever implies the alternative of "all or nothing"' (Derrida 1988: 116).[10] However, there are limits to this logical structure, and these limits are laid bare in the *aporia* that arises in the relation between foundational and anti-foundational thinking, in response to which an either/or or an all-or-nothing approach is inappropriate (Hurst 2004).

Hurst (249) maintains that the differential logic that characterises this *aporia* also gives rise to the following consequence: 'every foundation established will always already have been relatively damaged, corrupted, ruined, from the instant of its institution, without this being reason enough to give up on it.' What this means is that 'what *is*' (i.e. a phenomenon that is transcendentally constituted) can never coincide with the object that gives rise to this phenomenon (i.e. the noumenon). The

---

[9] See Sect. 2.5, feature 8.

[10] See Sect. 6.5.1 for the full citation.

transcendental constitution of a phenomenon requires that we suspend the aneconomic force of *différance*. However, this suspension of the aneconomic dimension destroys the play of *différance*, understood as something other than a regulated economic movement, which means that the thing 'itself' (the saturated fullness of its meaning) is lost. It has, in fact, always been lost, since the aneconomic dimension of *différance* also means that an object can never correspond with itself. Hence, every act of constitution is 'a false entry or a false exit' (to re-cite Derrida). Yet, the play of *différance* can also not be thought of only in terms of its aneconomic dimension, as this would imply that difference and distinction (as necessary conditions for the constitution of meaning) would be impossible, which in turn implies that phenomena would be unrecognisable. In other words, the economy of meaning would be unconstrained, and thus cease to be an economy.

In the happening of *différance*, Derrida therefore acknowledges transcendental or intentional constitution, as conceived of by Kant and Husserl, but also qualifies it by demonstrating how *différance* 'is self-annulling (its economic aspect is always already ruined by its aneconomic aspect)' (255). This qualification of transcendental constitution, whereby the constitution of phenomena also implies their relative ruin, is what is denoted by Derrida's *quasi-transcendental* or *limit* concepts of which justice, the gift, and ethics are examples. These examples serve to concretise the above account of meaning, and will be explored next.

3.4.2 Examples of Derrida's Double Logic

The nature of the play between the economic and the aneconomic, the restricted and the general, or calculation and excess characterises the double, *aporetic*, or quasi-transcendental logic that Derrida employs to explore the possibility of an opening in the restricted economy. Such an opening is premised on positing a relation between concepts and non-concepts, and is well-explained in Derrida's analysis on counter-violence or the other to violence, which is denoted by the concept of *gift* or *justice as gift*.

3.4.2.1 Violence and the Gift

In an excellent article, titled 'The time of violence: deconstruction and value', Elizabeth Grosz (1998) explores the theme of violence in Derrida work. Drawing on Derrida's (1976a) text 'The Violence of Writing' that appears in *Of Grammatology*, Grosz unpacks Derrida's three-pronged understanding of violence, defined as primordial, compensatory, and empirical violence.

Primordial violence, which is also denoted by the terms *arche-violence* or *arche-writing*, represents the 'original' splicing or cutting of the world. It is a violence that comes to the fore in the establishment of distinctions (for example, the first grunt for food distinguishes the grunt from non-grunts), which is also the manner in which phenomena (and therefore meaning) are constituted. This level of violence gives rise to, what Grosz (1998: 193) calls, 'the ontological equivocation of différance'. Compensatory or reparatory violence is aimed at erasing primordial violence, and 'designates itself as the moral counter of violence'. This form of violence is viewed as legitimate or sanctioned violence and is variously named law, right, or reason. Empirical violence is mundane or visceral violence, but Derrida (1976a: 112) contends that it 'is all the more complex in its structure because it refers at the same time to the two inferior levels of arche-violence and of law.' This is because it breaks open primordial violence to the extent that it iterates the form of this violence, but it also denudes compensatory violence, by issuing a challenge to that which has become naturalised, and that consequently functions 'as the proper, the so-called proper ... *perceived* by the *social* and *moral consciousness* as the proper, the reassuring seal of *self-identity*, the secret' (112). In other words, it opens up the system to the play of the aneconomic dimension of *différance*, which serves to destabilise our established, regulated, economic concepts.

The law or rights cannot therefore properly be called the counter to violence, even though it pretends to be such. This is because laws are always limited and exclusionary, and are therefore an example of compensatory or reparatory violence. What then is the counter to violence? The answer lies in that which exceeds calculation, and therefore meaning, and that can never be established in the world, except as an experience of the (im)possible.[11] One such an experience is denoted by the concept of the *gift*.[12]

In the conventional understanding, Derrida (1994: 11–12) notes that a gift implies that some "one" gives some "thing" to some "one other"'. Yet, Derrida (12) argues that '[t]hese conditions of possibility of the gift... designate simultaneously the conditions of the impossibility of the gift', to the extent that they 'produce the annulment, the annihilation, the destruction of the gift.' To understand this argument, consider the nature of a gift: a gift is always given within a restricted economy of exchange. It is recognisable as a gift, which in turn implies that receiving a gift

[11] As will become clear from the analysis (and as argued in Sect. 6.2), Derrida does not view *possibility* and *impossibility* as modal opposites. Rather, the condition of possibility is premised on (and hence tied to) the condition of impossibility. For this reason, he visually links the terms in his formulation of the *(im)possible*.

[12] It is interesting to note that the question of the gift also concerned Bataille, who, in his texts on economy, draws heavily on Marcel Mauss's work, titled *The Gift: The Form and Reason for Exchange in Archaic societies* ([1950] 1990). The opposition between the commodity and the gift in this work resonates strongly with Bataille's understanding of the opposition between the restricted and the general economy. According to Shershow (2001: 483), this 'moral opposition' essentially amounts to the difference 'between the economistic commodity and the generous gift.' For a detailed critique of this view of the gift see Shershow, especially Sects. 3 and 4. For a comparative exposition of the question of the gift in the works of Bataille and Heidegger (who influenced Derrida's understanding of the gift), see Comay (1990).

generates a debt to the giver. The gift is repaid by showing gratitude or reciprocating. Even declining a gift means that one has already acknowledged that a gift has been given, and the same debt is therefore implied. In order to give a true gift (which transcends the exchange of presents) the giver cannot know that she has given a gift, and the receiver cannot know that a gift has been given to her. In other words, a true gift can only be given in absolute secret, which is impossible. Hurst (2004: 252) explains as follows:

> If the gift has to be recognisable in order to be a gift, then the gift is impossible, for the recognition destroys it by converting it into a present. On the other hand, if the gift has to be unrecognisable in order to be a gift, then the gift is again impossible, for in the lack of recognition there is no gift.

Like *différance*, the gift is defined by both an economic dimension (the present) and an aneconomic dimension (the gift) (Hurst 2004), and therefore '[t]he gift is both part of and, in some sense, always beyond the economy of exchange' (Grosz 1998: 200). Yet, this does not mean that we ought to give up on gifts, only that we should approach gifts with a certain awareness. In Derrida's (1994: 30) words:

> *Know* still what giving *wants to say, know how to give,* know what you intend to give, know how the gift annuls itself, commit yourself [*engage-toi*] even if commitment is the destruction of the gift by the gift, give economy a chance.

To give a gift with awareness means to acknowledge the *aporetic* logic at play; which, as is the case with *différance*, implies that the condition that makes the gift possible is also the condition that makes the gift, understood in its full meaning, impossible. This awareness is precisely what allows us to 'give economy a chance', as it works to create an opening in, or transgression of, the restricted economy at the very moment that we, through our actions, also affirm the system. (This *aporetic* view of economy thus also provides us with an alternative to Luhmann's informationally-closed economy.) As such, Derrida (1992: 40) describes the movement of the gift as 'the becoming temporalization of temporalization, the animation of a neutral and homogeneous time by the desire of the gift and its restitution'.

For Derrida, justice is the rightful domain of the gift, and justice therefore presents us with the counter to violence. Unlike the law, which is both born from violence and which perpetuates a certain legitimate or sanctioned form of violence, justice as gift cannot be 'sent or received with a debt or structure of return'. As such, it gives time; 'it gives temporality, delay, a calculation of timeliness' (Grosz 1998: 200). Justice as gift breaks open the restricted economy of reciprocal exchange and creates the space for making a decision. In order to better understand this claim, it is useful to turn to the text, 'Force of Law: the "mystical foundation of authority"', in which Derrida (1992: 4) poses the following question: 'Does deconstruction insure, permit, authorize the possibility of justice?'. In answering this question, Derrida begins by unpacking his understanding of law and justice, and then reflects on the relation between these concepts. The gist of his argument is briefly summarised below.

3.4.2.2 The Relationship Between Law and Justice, Ethics and Politics

In the 'Afterword', Derrida (1988: 133) asserts that 'laws, symbolic inventions, or conventions [are] institutions that, in their normality as well as their normativity, entail something of the fictional'. He is quick to add that this does not make them fictions. However, on a structural level, they possess a certain inevitable fictionality since they are not natural givens. Derrida (1992) argues in 'Force of Law' that the ultimate foundation of law is by definition unfounded (hence, the characterisation of the basis of law as 'the "mystical foundation" of authority', which constitutes the subtitle of this essay). The law's foundation cannot be justified by an appeal to legal infrastructure. Rather, the law is enacted through virtue of a performative force which 'in itself is neither just nor unjust and that no justice and previous law with its founding anterior moment could guarantee or contradict or invalidate' (13). The paradox of law thus amounts to the fact that the founding distinction—what Derrida refers to in this text as 'interpretive violence' (13)—between what constitutes legal and illegal or just and unjust actions cannot itself be justified by an appeal to these categories. As such, we can challenge and deconstruct the law.

Justice, on the other hand, is something that necessarily lies beyond law and beyond our codified systems of meanings. This is because, in principle, justice should be able to account for full meaning. However, as Cilliers (2004: 23) notes:

> It is impossible to arrive at a complete and just description of society, not because we lack the resources, but because the demands made on such a description are contradictory. To provide justice to someone will mean that somebody else is treated unjustly. One cannot begin to think about the problem of justice if one does not accept its impossibility.

The impossibility of justice, or the inability to realise the fullness of justice because of the overdeterminacy[13] that defines our contingent contexts, means that justice as an ideal construct cannot manifest in the world. In other words, it lacks a concrete (political) position, it does not have reified meaning, and it is consequently undeconstructable. As such, justice is an ethical, as opposed to a legal, concept. The fact that justice lacks meaning does not mean that we should give up on the idea of justice, but it does require of us to rethink the relationship between justice and law in a manner that acknowledges the complexities involved. To this end, Derrida applies his quasi-transcendental logic to the question of justice and law in order to demonstrate that, firstly, a just decision is both regulated and unregulated (i.e. it demands both calculation and a grappling with the undecidable); and, secondly, the conditions that make a just decision possible also imply its relative ruin.

Derrida (1992) argues that justice is a singular event, whereas law is embodied in the generality of a rule, norm, or universal imperative. Given this, Derrida (17) asks: 'How are we to reconcile the act of justice... with rule, norm, value or the imperative of justice which necessarily has a general form, even if this generality prescribes a

[13] Derrida (1999: 79) remarks on the problem of overdeterminacy in writing that: 'I would say that the text is complicated, there are many meanings struggling with one another, there are tensions, there are equivocations; but this doesn't mean that there is indeterminacy. On the contrary, there is too much determinacy. That is the problem.'

singular application in each case?' In response to this question, Derrida again refers us to the *aporetic* structure that defines his logic: on the one hand, justice is responsibility without limits, and therefore it is necessarily excessive, incalculable, and before memory (even the very predicates of the responsibility that informs justice are open to deconstruction). On the other hand, the exercise of justice requires legitimacy or legality, stability and statute, calculation and codified prescription. The structure of law therefore ensures the possibility of deconstruction, and justice makes questions concerning law and the subject of rights possible. Hence, Derrida (15) writes that 'deconstruction takes place in the interval that separates the undeconstructibility of justice from the deconstructibility of *droit* (authority, legitimacy, and so on).' This implies that the decision between just and unjust is neither insured by a rule, but nor can it be the outcome of a purely private morality. The affirmation of justice therefore requires the position of law, even though justice can never correspond with law.

Derrida (28) confirms this interpretation in stating the following: '[t]hat justice exceeds law and calculation, that the unpresentable exceeds the determinable, cannot and should not serve as an alibi for staying out of juridico-political battles… And so incalculable justice *requires* us to calculate.' Derrida warns that if this were not the case, the incalculable idea of justice could be reappropriated by the most perverse calculation. In other words, without our legal fictions, justice would amount to little more than an expression of our private, relative moralities. He further argues that we should not only calculate, we should also strive to transgress the relation between the calculable and the incalculable in the direction of the other. Therefore, although it is impossible, law should strive to embody justice, as justice strives to gain expression in law. Yet, because the relation between justice and the law or the incalculable and the calculable is *aporetic* in nature, it also means that an irreconcilable tension defines this relation, which Hurst (2004: 257) formalises as follows: '[t]he very condition that makes a just decision possible (namely the preservation of law), is precisely the condition that simultaneously makes a genuinely just decision impossible (since justice requires the suspension of law).' Gayatri Spivak (1994: 19) summarises this dilemma as follows: 'responsibility annuls the call to which it seeks to respond.' Rather than leaving us feeling incapacitated, this dilemma should generate a sense of urgency, as Derrida (2002a) explains in 'Ethics and politics today'.

In this text, Derrida turns his attention to the relationship between ethics and politics, which is also defined by the *aporetic* logic that ties justice to law. Herein, Derrida writes that both ethical vigilance and political action are characterised by the greatest urgency. This is because both reflection and action demand a thoughtful and responsible response to the question 'what should I do?'. Furthermore, although our responses should be governed by infinite vigilance, such vigilance is always interrupted by the urgency of the situation. Therefore, to be ethically-politically responsible demands, on the one hand, a careful reflection on the excesses that cannot be accommodated in the utilitarian calculus of a conservative exchange economy; and, on the other hand, demands that we participate in this economy of calculation, evaluation, order, and judgement. In Derrida's (2002b: 25) words: 'One must not be content with affirmation. One needs position'. Derrida (2002a: 300) further elaborates by stating that '[t]he aporia here (or the double bind, if you prefer) is not the knowledge of whether one must choose the urgency over the non-urgency, rather it is always urgency against urgency'.

According to Derrida (2002a: 300), the aim, then, is 'to tie the most successful knot between two irreconcilable tensions'; or—as he states in another context—'to try to reduce a certain delay or gap between our political and theoretical deconstructions' (Derrida 1976b: 113). In other words, we should struggle on two fronts simultaneously: against institutions *and* the philosophical preconditions that make institutions possible, even if, in acting, we lose our radical position by becoming inextricably involved in the systems and oppositions that we wish to criticise[14] (Culler 1983).

The nature of urgency is therefore paradoxical and *aporetic*, since urgency is both the condition of the possibility and impossibility of all responsibility: without urgency one would have only the 'deployment consequent to a determinate knowledge' (298). This is because, as explained by Hurst (2004: 259):

> if infinite time and knowledge were at one's disposal, the very idea of the decision, and with it the very idea of justice would be annulled. Any decision made on the basis of such infinite knowledge, would not be a decision, but a mere calculation.

Hurst claims that urgency enables decision-making by freeing us from the prison of infinite time. Simultaneously, however, urgency renders the decision imperfect. Yet, the fact that decisions, actions, laws, and statutes are necessarily imperfect is not merely a practical problem concerning finite limitations. Rather, it is precisely because knowledge itself is characterised by an aporia (that is, by both an economic and an aneconomic dimension) that perfect knowledge is, in principle, impossible. It is thus the very *aporia* of meaning that inaugurates the *aporia* of ethics and politics, in that every action is a political move that serves to constitute phenonema, but also invites (indeed, demands) ethical reflection is so far as the act of constitution implies the relative ruin of these phenomena. And so, like a restless ghost, the demands of justice continue to haunt us from a place that is neither beyond nor within the world, always whispering the promise of the impossible gift.

3.4.3 *Tying the Knot Between the Restricted and the General Economy*

The above analysis of Derrida's double logic points to his deep and ongoing concern with the possibility of an intersection in our lived contexts between the codified meaning that characterises the restricted economy and the excess meaning generated by the sovereign operation. As demonstrated, the tension between the order of the calculable and the order of the incalculable (i.e. the complex) is never resolved

[14] This point is explored at length in Sect. 4.3.1.

within his work, but is instead marked by an *aporetic* relation, which requires of us to think together these irreconcilable orders.

The difficulty of thinking about the general economy hinges on the question of transgression, and although I shall not endeavour to trace the subtleties of Derrida's (1978a) argument in 'From restricted to general economy', it is nevertheless important to refer to his understanding of transgression (as it is explained in this text, as well as in Trahair's (2001) commentary on this text). As previously noted,[15] the sovereign operation (i.e. the immediate summit) cannot be said, except from the perspective of the spiritual summit (i.e. when it is already in decline). Sovereignty is therefore absolutely neutral and indifferent: it cannot be related to meaning and meaning can therefore also not be subordinated to it (since this would imply comparison). Following Bataille, Derrida (1978a: 264) writes that '[s]overeign silence… *tolerates no relations,* tolerates no symmetry with that which tilts and slides in order to be related to it'. How then must one think a sovereignty that cannot be said, but nevertheless demands to be said? In other words, how do we think on, and speak of, systemic openness? Bataille is sensitive to the limit condition of transgression, i.e. the manner in which transgression is related to that which it negates. For this reason, Bataille argues that we can only understand the movement of transgression as transgression, if it 'conserve[s] or confirm[s] that which it exceeds' (274). For Bataille, our understanding of transgression therefore operates under the Hegelian system (i.e. the system of meaning). Bataille (in Derrida 1978a: 275) describes dialectical sublation as follows:

> It is useless to insist upon the Hegelian character of this operation, which corresponds to the moment of dialectics expressed by the intranslatable German verb *Aufheben* (to surpass while maintaining).

Bataille's transgression therefore signifies a form of sublative negation (which, to recall, preserves and maintains that which is sublated). This is because any attempt at thinking about it necessarily reintroduces the metaphysics that he tries to avoid (in other words, any attempt to understand the system is already framed and delimited by the logic that is operative within the system).

Derrida (1978a: 275) explains this logic as follows:

> The Hegelian *Aufhebung* is produced entirely from within discourse, from within the system or the work of signification. A determination is negated and conserved in another determination which reveals the truth of the former... The *Aufhebung* is included *within* the circle of absolute knowledge, never exceeds its closure, never suspends the totality of discourse, work, meaning, law, etc.

Derrida however questions Bataille's resigned attitude to transgression's inevitable return to the economy of meaning. Although he acknowledges the Hegelian character of transgression, he nevertheless argues that there exists a fundamental difference between dialectical sublation and sovereign transgression (Trahair 2001). For Derrida, transgression cannot be entirely maintained in Hegel's restricted economy. Rather, 'the transgressive relationship… links the world of meaning to the

---

[15] See Sect. 3.2.2.

world of nonmeaning' and therefore 'designate[s] a movement which properly constitutes the excess of every possible philosopheme' (275). As explained by Trahair (2001), sovereign transgression therefore utilises the noumenal form (what Derrida calls, the *empty* form) of the Aufhebung in an analogous fashion in order to *relate* the restricted and general economies to each other. Bataille is correct in thinking that transgression needs the restricted economy of determinate meaning, but he is incorrect in thinking that transgression cannot be experienced except when sublated to meaning. Rather, from a Derridean perspective, the 'transgressive affirmation of the general economic operation' (167), in laughter for example, has the implication that every restricted economy is already open to the effects of the general economy. This means that the play of meaning cannot be arrested. Sovereignty therefore interiorises lordship, or laughter interiorises the *Aufhebung*, rather than the other way around.

In 'Structure, sign, and play', Derrida (1978b) again affirms the interpretation that there is no pure restricted economy in speaking of our phenomenological engagement with the world. In this text, he writes that, contrary to Hegel's vision, totalisation is impossible due to our very ontology. This is because:

> the nature of the field... excludes totalization. This field is in effect that of *play*, that is to say, a field of infinite substitution only because it is finite, that is to say, because instead of being an inexhaustible field, as in the classical hypothesis, instead of being too large, there is something missing from it: a centre which arrests and grounds the play of substitutions (289).

To treat meaning only in terms of the utilitarian analysis of a restricted economy is to deny the very nature of the field, which is characterised by the absence of a central organising principle. In this regard, Derrida (1978a) argues that even the Hegelian economy is vulnerable to the iterative play of meaning. This means that reductive strategies will necessarily fail. Instead, we should recognise the need to grapple with the excesses or complexities, even though we cannot express them in the language of concepts. Alternatively, to speak of two unrelated economies—one restricted and meaningful and the other general and excessive—as does Bataille, is to deny the productive affirmation of the movement of the general economy within the restricted economy. As demonstrated in this section, what Derrida requires of us is to take cognisance of the limits and constraints that govern our general theory or understanding of the economy, and to transgress and supplement this understanding by what Derrida (1988: 117) calls 'another "logic" that accounts for the impossibility of concluding such a "general theory"'. Moreover, this task 'requires endurance in thinking' whereby we resist the desire to halt at the *aporia* or to overcome it (Derrida 1993). In other words, it requires that we engage in the *aporetic* logic that characterises meaning.

The above reference to endurance recalls Nietzsche's injunction to become who we are, which requires both that we hold a multitude of perspectives in place without crumbling under the contradictions that exist between these various power complexes, and that we affirm the play of the world and resist the temptation of assigning a higher purpose or meaning to our lives. Although, Nietzsche's project certainly constitutes a positive affirmation of the general economy, his desire to understand—and hence, in a certain sense, to transcend—this general economy, ultimately leads to a unproductive understanding of economy. In an unpublished note, Nietzsche writes: '*Most difficult to unite:* a will, strength of the fundamental feeling and change of movement (changes)' (Nietzsche 1999b: 97; trans. van Tongeren). It is specifically with regard to this point on engagement and disengagement that Wolfgang Müller-Lauter (1971) argues that the failure of the Nietzschean project becomes evident. This is because 'the will to power is torn between the strong enforcement of some particular position on the one hand and the relativizing of every position as perspectival on the other' (in van Tongeren 2000: 237). The difficulty here lies in trying to reconcile these contradictory perspectives (in the fictional figure of the *Übermensch*), rather than to maintain and to work with the tensions between these two positions. These positions denote the struggle between the controllable, determinable, and hierarchical aspect of the restricted economy on the one hand, and the indeterminable and excessive aspect of the general economy on the other. In complexity terms, it denotes the struggle between meaningfully engaging with complex systems through modelling (and thereby through reducing the complexity), whilst simultaneously acknowledging the excesses generated by complexity that govern our lived experiences (and that threaten the very integrity of our models).

At this juncture, it should be clear that the restricted and general dimensions of meaning—or the relation between models of complex phenomena and complexity itself—are better understood in terms of *aporia* than unification. As concerns our understanding of economy, this point elevates Derrida's work not only above Bataille's dual notion of economy, but also above Nietzsche's attempt at mastering economy.

3.5 Nancy

Before teasing out the implications that this analysis of economy holds for our understanding of complex systems, I wish to explore one further position, namely that of Nancy's work on community and thinking. Despite the productive insights that have come to light in the investigation thus far, the question of economy has essentially been 'governed by the theme of the sovereignty of a *subject*' (Nancy 1991: 23). Indeed, Nancy (24) notes that despite Bataille's interest in community, and although Bataille 'had no *concept* of subject… he allowed the communication exceeding the subject to relate back to a subject, or to institute itself as subject.' The same can be said of Derrida in that the endless chain of signification that keeps meaning in play, does not do away with the subject, but complexifies the question of the subject and her relation to her life-world. Nancy however approaches the question of meaning (and therefore of economy) from the perspective of community, which generates a different understanding of systemic openness.

Nancy's focus on community is especially relevant in the context of this study, since it allows him to define the subject as an *a priori* relational being (we are constituted as singular-plural), rather than viewing relationality primarily as the consequence of the process of signification. Nancy further takes a radically immanent view of the question of existence (as expressed in community), which, as will be demonstrated, bears interesting consequences for our understanding of systemic openness.

3.5.1 On Community

Nancy (1991: 23) notes that '[f]or Bataille, as for us all, a thinking of the subject thwarts a thinking of community', yet he believes that 'Bataille thought nothing else but this very thing he gave up thinking' (25). Therefore, Nancy states that his reading of Bataille should not be viewed as a critique or a reservation, but rather as his own engagement with Bataille's experience of the limit of thought, which for him is denoted by the ideas of community and communication.

Although Bataille did not pursue his thoughts on community, Nancy argues that he is the first thinker to acutely experience modern community as a space, rather than a lost communion or a project. Moreover, community as a space constitutes 'a spacing of the experience of the outside, of the outside-of-self'[16] (19). Community is not something that can be thought of as separate from finite beings, but as the very condition for a being's constitution. The nature of being is therefore that of being-in-common because 'singular beings… appear only to the extent that… they are exposed, presented or offered to one another' (58). Otherwise expressed, singular beings '*compear*' (28) in the sense that '*you share me*' (29).[17]

This recognition of our compearance results in what Nancy calls *clear consciousness*. He defines this as 'Hegelian self-consciousness… *suspended* on the limit of its access to *self*' for the reason that 'consciousness *of* self turns out to be outside the self of consciousness' (19). Clear consciousness interrupts self-consciousness, and is what takes place in community. In Nancy's (19) words, '"clear" consciousness… can only take place as the communication of community: both as what communicates

[16] Nancy makes extensive use of hyphenated compounds, either by inserting hyphens in words that are usually unhyphenated (e.g. ex-posed, im-mediation) or by constructing hyphenated phrases (e.g. outside-of-self). In French the term for hyphen is *trait d'union,* which literally translates as 'line or mark of union'. Bearing this in mind, Peggy Kamuf (2005: 278) summarises Nancy's reasons for using the hyphen as follows: On the one hand, the hyphen represents 'the graphic imperative… [the] necessity of going over the articulating trait that has been covered or closed down.' On the other hand, the hyphen as 'graphic practice is thematized… around the preposition "with" or "avec", which serves as the mark of union/disunion, and which marks the interval between/among substances.'

[17] See Sects. 5.4.1 and 5.4.2 for a detailed discussion on Nancy's understanding of being as singular-plural and in community with others.

within community, and as what community communicates' (19). Communication in community (as an expression of clear consciousness) defines our ontological condition; it is *'an experience that makes us be'* (26). Therefore, although communication elicits the desire for Hegelian recognition, there is something more fundamental at play in the act of communication, namely 'knowing without knowledge, and without "consciousness," that *I* am first of all exposed to the other, and exposed to the exposure of the other' (31).

Following Bataille, Nancy further argues that communication itself characterises the limit, in that it is only through communication that we can express sovereignty at a moment that it is nevertheless in decline. As such, Nancy (26) writes that 'only a discourse of community, exhausting itself, can indicate to the community the sovereignty of its sharing (that is to say *neither present* to it nor *signify* to it its communion).' Our ontological condition is therefore one in which the limit of our finitude is exposed (as opposed to sublated), and community is nothing other than this exposition. Relating the analysis back to our understanding of economy, one can argue, as does Shershow (2001: 490), that 'community is formed by *finite* beings who are *infinitely* ex-posed to one another; and also by the infinite ex-position of finitude.' In offering ourselves up to communication, Shershow (491) writes that 'the restricted economy relates to the general economy *not* as its truth, and not as the voice of some inescapable practicality that belittles theory; but rather as its shared condition of possibility.' To better understand this claim it is useful to turn to Nancy's analysis of thinking, as detailed in part 1 of *A Finite Thinking* (2003).

3.5.2 *On Thinking*

The workings of the general economy—i.e. the manner in which we are infinitely exposed to one another and to our own finitude—are felt due to the brute fact of our finite existence, in which sense circles endlessly, and where no outside is possible. Benjamin Hutchens (2005: 44) explains that, for Nancy:

> the world does not "have" an exterior sense… but sense "belongs to its structure" as a constitutive "signifyingness" or "significance" of the world itself. The "there is", the happening of experience to existence itself, is the "constitutive" sense itself.

Nancy supports an open immanence, which he views as the only viable response to the end of philosophy or of transcendence. For Nancy (2003c), the death of God is something that we note, and think after. That, he maintains, is all. Thinking that takes place at the limit of thought, necessarily implies a grappling with sense and with our finitude. Thinking does not exhaust or denounce sense; rather, it is the endless act of trying to make sense of sense. Nancy (27) gives a succinct overview of what thinking entails in his essay 'A finite thinking', and it is worth citing the relevant passage at length. A finite thinking he maintains is:

> Not a thinking of relativity, which implies the Absolute, but thinking of *absolute finitude*: absolutely detached from all infinite and senseless completion of achievement.

> Not a thinking of the limitation, which implies the unlimitedness of a beyond, but a thinking of the limit as that on which, infinitely finite, existence arises and to which it is exposed.
>
> Not a thinking of the abyss and of nothingness, but a thinking of the un-grounding of being: of this "being," the only one, whose *existence* exhausts all its substances and all its possibility.

In short, a finite thinking is non-teleological in that thinking, like being, has no higher or external purpose. The proper object of thinking is therefore not a symbolic-ideological order but rather the order of infinitely finite existence itself. The goal is therefore to 'think anew and with all rigour the inaccessibility of sense as the very means of accessing sense' (14). Indeed, a 'finite thinking', Nancy (14) contends, 'is one that *rests* on this im-mediation.'

In an essay titled 'Concealed thinking', Nancy (2003a) again engages with Bataille, who, along with Sartre, he credits as having understood the nature of thinking as necessarily transgressive, incomplete, and excessive, and hence concealed from itself. Nancy (35) writes 'that both Sartre and Bataille... are anxious [*dans l'angoisse*], experiencing the cessation of a sense that is neither a lack nor a loss but the point at which truth arises as this very cessation.' In other words, both understand thinking as something other than the contemplation of sense understood in substantive terms. Nancy (36) agrees with this view of thinking, and identifies the important challenge in this regard as 'one of knowing (or of not-knowing...) how to think a thinking that is *still* a thinking even when its content is not-knowing', or, otherwise expressed, how to grasp thinking 'without ever capturing or reducing [it] to something that has been caught' (46). (This challenge constitutes yet another phrasing of the central problematic of this chapter.)

In addressing this challenge, Nancy refers to Bataille's understanding of 'not-knowing', which he interprets as follows:

> To *not-know* [is] to enter into the obscurity and the opacity of what is no longer a matter of knowing in any way, shape, or form. The thinking conceived thus is "still a thinking," then, but in a sense hitherto unknown. It introduces a change in level and a rupture in thinking itself: it is thinking concealing itself from itself (37).

Nancy also draws on Bataille's metaphor of the night in which we see and experience obscurity *as* the privation of sight in order to explain this understanding of a thinking whose content is not-knowing or of a sense whose sense is that of self-concealment. The night becomes a metaphor for truth. This truth does not constitute an object of knowing 'divined from its contours, its breath, its rustling' (38), but is rather 'the element of invisibility' (38) that defines truth in the sovereign sense. If thinking is equated with sight, concealed thinking (i.e. the contemplation of the night of truth) is 'representation as the sight of nothing rather than a nothingness of sight' (37). Since there is nothing to see, the power of sight is stretched to its limit. In this regard, the truth of sense, which is one of self-concealment, '*touches* on its extremity as the eye touches on the night in which it is lost. Self-possession shows itself to be outside of itself' (39). This formulation echoes his view of community, which he describes in *The Inoperative Community* as 'the ecstatic consciousness of the night of immanence, insofar as such a consciousness is the interruption of self-consciousness' (Nancy 1991: 19).

The link between thinking and community (or more specifically, community in communication) is made explicit in 'Concealed thinking', where Nancy (2003a) traces the etymology of the word *to conceal* (*dérober*), and explores its association with disrobing or denuding. For Nancy, a concealed thinking is necessarily a naked thinking, i.e. a truth. And, it is this nakedness or truth that we share or have in-common, in that—as he explains—'[t]here is no solitary nakedness. If I am naked and alone, I am already an other to myself, an other with myself' (39). He further writes that '[c]oncealed thinking is identical to communication, and this identity is itself the night of not-knowing' (45).

Thinking as the night of not-knowing further serves to draw attention to the ungrounding of being, or to the fact that we are non-identical to ourselves (as typified in *The Inoperative Community* by the term *clear consciousness*). For Nancy, this is however not a paralysing thought but the very grounds for praxis, which he understands as 'the endless transformation of the subject of sense in itself' (47), and which is none other than a subject of communication and concealment.[18] He thus concludes 'Concealed thinking', stating that:

> The concealment of thinking is its praxis: thinking that undoes its objects in order to become the thinking that it is: *we*, with one another and with the world (47).

3.5.3 Nancy's Immanence

Of all the thinkers analysed in this chapter, Nancy's position is the most radically immanent. In his essay, 'The Unsacrificable', Nancy (2003b) interprets Bataille's sovereignty as the finite condition of our existence. This is because if sovereignty is nothing (as Bataille claims), then neither is existence in that the essence of existence and of being is existence itself. Otherwise expressed, existence has no essence and is therefore nothing more than 'the sense of being as finitude of sense' (74). This also means that nothing precedes or follows the happening of existence; in Nancy's (75) words: '*it isn't even offered or sacrificed to a Nothingess, to a Nothing or to an Other, in whose abyss it could still impossibly enjoy its own impossibility of being.*' This reference to the impossible recalls Derrida's description of our experience of the gift of justice. Derrida's work constitutes a challenge to the 'seemingly transcendental difference between inside and outside… [as] a 'permanent requirement' of thought' (Lucy 2004: 53). However, when compared to Nancy, it becomes evident that Derrida still works with the difference between interiority and exteriority to the extent that he affirms the existence of an alterity that cannot be assimiliated (for him, the alterity of justice becomes the transcendental condition of otherness[19]). In

[18] See Sect. 6.3 for a more detailed discussion on Nancy's praxis.

[19] Wessel Stoker (2012: 20) classifies Derrida's metaphysics as 'transcendence as alterity'. In this view, '[t]he relationship between transcendence and immanence is no longer viewed as an opposition. Rather, one has learned to think beyond the opposition, whereby the wholly other can appear in every other.' In contrast to Derrida's position, Nancy argues that the wholly other appears in me,

his total affirmation of immanence, Nancy's position is also more radical than Bataille's, who, he argues, fails to renounce sacrifice, and hence metaphysical form (even though this form is only retained negatively, as a lack). This is because—according to Nancy—sacrifice 'is always connected to the fascination with an ecstasy turned toward an absolute other or toward an absolute Outside' (Nancy 2003b: 75).

Existence is therefore the experience of sovereignty in as much as the sovereign is neutral, indifferent, or nothing. Nancy writes that existence grants nothing (there is nothing but the event of existence). 'Nothing' should therefore not be thought of as that which is exposed to an outside; rather:

> "Nothing" affirms finitude and thus "nothing" immediately leads back to itself and to nothing else. It de-subjectifies it, removing from it any possibility of its being appropriated by anything other than its own event, its advent. This sense of existence, its sense proper, is unsacrificable (75–76).

For Nancy, this radical immanence does not inspire the horror of the Eternal Return. It does not require the fortitude of the *Übermensch*, and it does not carry the consequence of fatalism. Rather, and as noted above, the finitude of existence is exactly that which enables the endless transformation of the subject of sense through thought. As such, Nancy (2003d: 247) coins the expression that *thinking is love*,[20] which can be contrasted with the Nietzschean injunction to love our fate (*amor fati*).[21] A being without essence signifies an 'ontological affirmation', which Nancy relates to freedom in that it expresses our attempts to make sense 'of the infinite absence of appropriable sense' (Nancy 2003b: 76). Since there is no outside, any attempt to appropriate sense becomes immanent; it becomes the horizon of our experiences. Nancy (76) describes this horizon as follows:

> The horizon holds existence at a distance from itself, in the separation or the "between" that constitutes it: *between* life and death... We don't enter into this between... Not because it would be an abyss, an altar, or an impenetrable heart, but because it is nothing other than the limit of finitude. And, this limit, if we're not going to confuse it with a "finiteness," Hegelian, for example, is a limit that leaps over nothing. Existence alone leaps, leaping over itself.

Nancy writes that, strictly speaking, there can be no horizon as there is nothing to transgress, i.e. no beyond to the experience of our existence. Yet, in another sense, horizontality—the quest to make sense from non-sense—is all we have: '[o]n the horizon something is constantly rising and setting' (76). Existence, then, is constrained or restricted by itself (there is nothing beyond existence); it is consti-

---

to the extent that I am defined as singular-plural in that my ontological condition is that of being exposed to ek-sistence or community.

[20] In this essay, 'Shattered love', Nancy (2003d) argues that the formulation 'thinking is love... can... begin the quest for an ignored essence of thinking for which we lack any evident access' (247). This is because framing thinking in terms of love draws attention to both our preoccupation with thinking (in philosophy, for example) and the elusive character of thinking. For a fuller description of Nancy's view of the relation between thinking and love, see Sect. 7.4.2.

[21] See Sect. 3.3.1.

tuted by beings-in-relation who are infinitely exposed to one another and to their own finitude (which constitutes the experience of sovereignty).

In complexity terminology, it is the *a priori* relational nature of systemic components (which means that relationality should be conceived of as a necessary condition for the existence of components themselves) that gives rise to systemic openness. For Nancy, systemic openness should not be thought of in relation to the boundary line which facilitates a differential relation between a system and its environment, as Derrida would have it; but as a boundary line, which is no more or no less than the sum total of our engagement with the system (i.e. immanence) itself. A complex system, in other words, is nothing other than the boundary or horizon constituted by components in relation with themselves and with one another. Cilliers (2001: 142) argues that 'in a critically organised system we are never far away from the boundary'. This is because the richly interconnected nature of a complex system means that 'there will always be a short route from any component to the "outside" of the system.' He further argues that since the boundary is always folded in, it is possibly the case that 'the system consists of boundaries only.' Nancy radicalises this point, since if there is no outside to the system, the idea of the inside of the system also falls away. All we have is the horizon of community, which is neither related to an interior nor to an exterior, but which constitutes our experience with, and our communication of, immanence itself.

Whereas Bataille views the general economy as separate to the restricted economy, and Derrida relies on his *aporetic* logic to relate the restricted and general economy to each other, for Nancy the restricted economy is nothing other than the general economy. It is the medium through which sense circulates. Bataille and—to a lesser extent, Derrida—continues to affirm the possibility of a beyond to meaning (as expressed in Bataille's dual notion of economy and in Derrida's *aporetic* or quasi-transcendental logic). For Nancy however there can be no domain beyond the world—not even in thought—and as such transcendence is impossible. It is, instead, *due* to the restricted nature of the economy of our existence that we are able to experience the concealment and truth of thinking, described as 'the night of not-knowing'.

3.6 Analysis

In *On Complexity*, Morin (2008) argues that—given the acknowledgement of complexity—there can be no omniscient vantage point. However, we can still avoid relativism by constructing meta-viewpoints, however fragile or limited they may be (92). In this regard, a theory of economy serves as such a meta-point of view, in that it helps us to think productively about the difficult issue of framing complexity. As mentioned in the introduction, the term *economy* holds two advantages, namely it is a relational concept, and it denotes systemic openness. In this penultimate section of the chapter, the main insights that were derived from the analysis of several different philosophical perspectives will be related back to these two strengths in order to further develop our understanding of the economy of complexity.

Four different notions of economy came to the fore in this chapter; namely Hegel's totalising economy, Nietzsche's and Bataille's dual economy, Derrida's *aporetic* economy, and Nancy's immanent economy. All four positions hold different implications for systemic relationality and openness, and these implications are summarised below.

Hegel's totalising economy constitutes a closed system in that philosophical reasoning is all-encompassing, and moreover is the medium through which the grand synthesis (i.e. the full presence of meaning) can be realised. Although Hegel's work on negativity has been instrumental in developing a dialectical understanding of meaning, his historical and teleological accounts of meaning nevertheless mean that all systemic resources are conserved, and put to work towards a definite goal. Systemic components only relate to one another in a dialectical, as opposed to a complex fashion. There is therefore also no opening in Hegel's system (even death is sublated and interiorised). This means that his meta-view is inadequate for understanding the economy of complexity.

In comparing Hegel's economy to the other positions developed in this chapter, we find support for Human and Cilliers' (2013: 40) claim that 'conservative positions are able to be progressive in appearance, as they maintain a faith in science and the possibilities of the future, yet remain tethered to the possibilities of the present economy.' In other words, true transgression is impossible from the vantage point of a closed system. However, since systemic resources are conserved in a closed system, such a system can give the impression that systematic and controlled knowledge generation is possible, and that the system is progressive.

Both Nietzsche and Bataille pioneered an understanding of economy, aimed at transgressing the utilitarian forces that govern the possibility of the present economy, by emphasising the aneconomic dimension (or the excess of meaning). Nietzsche's focus on the force-field of power wills that stand in a dynamic relation to one another and that constitute our world, serve to break through the rigid constraints of an exchange economy. Yet, Nietzsche's desire to come to terms with the entire power field betrays his desire to master complexity in all its complexity. Although he foresaw the impossibility of achieving this, he nevertheless postulates a domain beyond the world in which this ideal can be recognised. Similarly, Bataille's preoccupation with the question of sovereignty—as the total loss of meaning and self—means that both these thinkers ultimately endorse a dual notion of economy. Such an economy is constituted by two realms: the economic (or codified and restricted) realm and the aneconomic (or complex and general) realm. This dual notion of economy highlights the difficulty of thinking about systemic openness, but also points to the impotence of their respective philosophical visions. We cannot develop a productive meta-view of the economy of complexity if complexity becomes either that which we can say nothing about from our limited vantage points or something which is simply reduced to these vantage points. In other words, what is needed is to relate complexity to meaning in order to develop a productive reading of the economy of complexity.

Derrida's *aporetic*, as opposed to dual, notion of economy attempts to achieve this aim. Derrida's starting point concerns the problematic of a system, described by Morin (2008: 100) as follows:

> The system cannot be grasped as pure unity nor as absolute identity, nor as decomposable composite. We need a systemic concept that expresses at once unity, multiplicity, totality, diversity, organization, and complexity.

Through means of his *aporetic* logic, Derrida tries to think together the concept of the system and the irreducible complex conditions that give rise to (but are not exhausted by) this concept. Derrida's quasi-transcendental logic expresses the manner in which phenomena (i.e. systems of meaning) are constituted, but also ruined in their constitution, since phenomena can never correspond with the ideal. Although Kant also recognised this point, Derrida radicalises it by focusing on the process of signification which gives rise to an excess of meaning. This excess of meaning not only prevents concepts from corresponding with the noumenon (Kant) but also destroys the internal integrity of these concepts (Hurst 2004). In his work, Derrida demonstrates that all conceptual language is open to the aneconomic force of *différance*, which means that our systems of meaning are necessarily open (and even open to an absolute collapse). Derrida concentrates his analysis around the idea of alterity (i.e. that which cannot be appropriated) in order to draw attention to the fact that meaning is always limited and exclusionary, and to *aporetically* relate the inside of the system to its outside (i.e. the alterity of the general economy). In so doing, Derrida successfully articulates the workings of the economy of complexity in emphasising the openness of all systems of meaning, as well as the relational character of meaning. As such, Morin's (2008: 109) self-description of his work cited below also rings true of Derrida's aims:

> It is not a question here of a Hegelian ambition to dominate the world of systems with the System of Ideas. It is a question of an inquiry into the relation, both hidden and extraordinary, between the organization of knowledge and the knowledge of organization.

Nancy develops the notion of the economy of complexity along different lines to Derrida. Nancy's immanent economy is premised on the idea that existence is community in communication, and nothing more. Otherwise stated: existence is organised relationality. Although on the above reading, Morin's views on complex systems resonate strongly with Derrida's understanding, Morin's (100) description of complex systems cited below seems to be more closely aligned to Nancy's view:

> Objects give way to systems. Instead of essences and substances, organization: instead of simple and elementary units, complex unities; instead of aggregates forming bodies, systems of systems. The object is no longer a form-essence or matter-substance. There is no longer a form-mould that sculpts the identity of the object from the outside. The idea of form is preserved, yet transformed; form is the totality of organized complex unity that manifests itself phenomenally as whole in time and space... Form is no longer conceived in terms of essence but in terms of existence and organization.

In the above citation, Morin (like Nancy) challenges the idea of an inside and an outside, and replaces it with a form conceived of only in terms of existence and organisation. However, by drawing a distinction between realism (the physical conditions for formation and existence) and formalism (the abstractions produced by the mind of the conceptualiser or observer), Morin nevertheless distinguishes between the system (the inside) and its environment (the outside), even though he argues that our understanding of a system must necessarily relate these terms to each other 'in a manner that is at once complementary, concurrent, and antagonistic' (105). Unlike Derrida, who views the outside of the system in terms of a transcendental alterity, Morin therefore frames the systemic outside in terms of a physical reality (in Kant's terminology, a non-appropriable noumenon). Nancy's preoccupation with immanence allows for a more radical interpretation of the above citation—one which truly does away with the difference between inside and outside.

Although Nancy probably would not deny the distinction between realism and formalism (as used by Kant and Morin), the value of his analysis does not centre on thinking the relation between these dimensions. Rather, Nancy construes existence in terms of relationality (or being-in-community). In his analysis, the system is not open towards an outside, but systemic openness is created by virtue of the fact that we are ex-posed to one another. Otherwise stated, the meaning created through communication in a finite community is excessive, and the impetus for communication is our very finitude. On my interpretation, Nancy does away with the distinction between the inside and the outside of the system by radicalising relationality. Not only is there nothing beyond our existence, but if being is *a priori* defined as singular-plural, then the difference between self and other can also no longer be thought of in terms of an inside or outside. Nancy thus views complexity as a consequence of relationality and his work represents a poignant example of economy, since meaning is constituted by the limit (or the horizon) which is the outcome of beings in relation.

3.7 Conclusion

The aim of this chapter was to think through the notion of organised relationality. Of specific interest was how the interactions between systemic components give rise to both codified meaning and an excess that cannot be articulated in meaningful terms. As such, both the nature of the relations between systemic components, as well as the nature of the relation between meaning and non-meaning, or codified knowledge and complexity, were explored.

Bataille is credited for highlighting the fact that every economy exceeds its own limits, and hence for recognising that the interesting question does not concern the relational dynamics of a limited or restricted economy (i.e. production) but rather the excess generated by these relational dynamics (i.e. surplus and waste). However, the manner in which he conceives of the relation between the restricted and general economies is, as previously argued, not very productive. On his count, such a relation amounts to a non-relation (the restricted and general economies are essentially conceptualised as two separate economies). Bataille's position can therefore be summarised as follows: (1) Complexity is generated by a constrained set of relations, which gives rise to a system (the restricted economy of codified knowledge) but also to excess. (2) This excess forms part of a system's environment (the general economy). 3) The environment cannot be understood from the perspective of the system.

Although Derrida and Nancy agree with the first two statements, both the work of Derrida and of Nancy represent a more fruitful engagement with the question of how the restricted economy relates to a general economy. Derrida still works with the classic distinction between the inside and outside of the system, but he relates the inside to the outside by framing these relations in terms on an *aporia* (which serves to emplace the outside within the system itself). Through means of the *aporia* Derrida is able to offer an account of systems as both bounded and open (which serves to render every conceptualisation as *a priori* ruined). Nancy foregoes the distinction between inside and outside in advocating a radically immanent stance. Complexity, for him, is contingent upon the radicalisation of being. As such, relationality is defined as an *a priori* condition of being. The inability of being to correspond to itself (we are singular-plural by nature) gives rise to complexity. Complexity manifests in the space between singularity-alterity, and is felt in communion with others on the horizon or limit, defined as existence.

Although, ontologically speaking, Nancy's and Derrida's projects differ substantially, their respective projects share an important implication for our phenomenological experience: namely that complexity, which manifests as the inability of meaning to be exhausted and to correspond with itself, marks the heart of our human condition. It will be argued in the following chapters that this fact alone necessitates—what I have termed—an ethics of living, which is conceptualised in terms of a sustained and serious engagement with complexity and the implications that it holds for knowledge generation, identity formation, agency, and action. These themes will be treated in subsequent chapters, and will also be explored at the hand of the example of foreignness in an attempt to concretise the theoretical insights.

References

Badiou, A. 2001. Who is Nietzsche? *Pli* 11: 1–11.

Bailey Gill, C. 1997. Bataille and the question of presence. *Parallax* 4: 89–98.

Bataille, G. 1970. Sur Nietzsche. In *Oeuvres Complètes*, vol. VI. Paris: Gallimard.

———. 1984. The notion of expenditure. In *Visions of excess: Selected writings*, ed. A. Stoekl, trans. A. Stoekl, C.R. Lovitt, and D.M. Leslie Jr., 116–129. Minneapolis: University of Minnesota Press.

———. 1991. *The accursed shared: an essay on general economy*, vol. 1, *Consumption*. Trans. R. Hurley. New York: Zone Books.

———. 1992. *On Nietzsche*. Trans. B. Boone. New York: Paragon House.

Baudrillard, J. 1987. When Bataille attacked the metaphysical principle of economy. Trans. D.J. Miller. *Canadian Journal of Political and Social Theory* 11: 57–62.

Bennington, G. 1995. *Bataille: Writing the sacred*, ed. C. Bailey Gill. London: Routledge.

Cilliers, P. 2001. Boundaries, hierarchies and networks in complex systems. *International Journal of Innovation Management* 5(2): 135–147.

———. 2004. Complexity, ethics and justice. *Humanistiek* 19: 19–25.

Comay, R. 1990. Gifts without presents: economies of "experience" in Bataille and Heidegger. *Yale French Studies* 78: 66–89.

Culler, J. 1983. *On deconstruction: Theory and criticism after structuralism*. London: Routledge.

Derrida, J. 1976a. *Of grammatology*. Trans. G. Spivak. Baltimore: Johns Hopkins University Press.

———. 1976b. Entre crochet (An interview). *Digraphe* 8: 97–114.

———. 1978a. From restricted to general economy: A Hegelianism without reserve. In *Writing and difference*. Trans. A. Bass, 251–277. London: Routledge.

———. 1978b. Structure, sign, and play in the discourse of the human sciences. In *Writing and difference*. Trans. A. Bass, 278–294. London: Routledge.

———. 1981. *Positions*. Trans. A. Bass. Chicago: University of Chicago Press.

———. 1982. Différance. In *Margins of philosophy*. Trans. A. Bass, 1–28. Chicago: University of Chicago Press.

———. 1988. Afterword. In *Limited Inc*, ed. G. Graff, trans. S. Weber, 111–160. Evanston: Northern Western University Press.

———. 1992. Force of law: The "mystical foundation of authority". Trans. M. Quaintance. In *Deconstruction and the possibility of justice*, ed. D. Cornell, M. Rosenfeld, and D.G. Carlson, 3–67. New York: Routledge.

———. 1993. *Aporias*. Trans. T. Dutoit. Stanford: Stanford University Press.

———. 1994. *Given time: 1. Counterfeit money*. Trans. P. Kamuf. Chicago: University of Chicago Press.

———. 1999. Hospitality, justice and responsibility: a dialogue with Jacques Derrida. In *Questioning Ethics: Contemporary Debates in Philosophy*, ed. R. Kearney and M. Dooley, 65–68. London/New York: Routledge.

———. 2002a. Ethics and politics today. In *Negotiations: interventions and interviews, 1971–2001*, ed. and trans. E. Rottenberg, 295–314. Stanford: Stanford University Press.

———. 2002b. Negotiations. In *Negotiations: Interventions and interviews, 1971–2001*, ed. and trans. E. Rottenberg, 11–40. Stanford: Stanford University Press.

Fink, E. 2003. *Nietzsche's philosophy*. Trans. G. Richter. London: Continuum.

Foucault, M. 1974. A preface to transgression. In *Language, counter-memory, practice. Selected essays and interviews*, ed. and trans. D.F. Bouchard and S. Sparks. Ithaca, NY: Cornell.

Gillespie, M.A. 1999. Nietzsche and the anthropology of nihilism. *Nietzsche-Studien* 28: 141–155.

Grosz, E. 1998. The time of violence: Deconstruction and value. *Cultural Values* 2(2/3): 190–205.

Habermas, J., and F. Lawrence. 1984. The French path to postmodernity: Bataille between eroticism and general economics. *New German Critique* 33: 79–102.

Hegel, G.W.F. 1977. *Phenomenology of spirit*. 2nd ed. Trans. A.V. Miller. Oxford: Clarendon Press.

Human, O., and P. Cilliers. 2013. Towards an economy of complexity: Derrida, Morin and Bataille. *Theory Culture Society* 30(5): 24–44.

Hurst, A. 2004. Derrida's quasi-transcendental thinking. *South African Journal of Philosophy* 23(3): 244–266.

Hutchens, B.C. 2005. *Jean-Luc Nancy and the future of philosophy*. Chesham: Acumen.

Kamuf, P. 2005. *Book of addresses*. Stanford: Stanford University Press.

Kant, I. 1929. *Critique of pure reason*. 2nd ed. Trans. N. Kemp Smith. London: Macmillan Press.

Kohn, M. 2005. Frederick Douglass's master-slave dialectic. *The Journal of Politics* 67(2): 497–514.

Kojève, A. 1969. *Introduction to the reading of Hegel: Lectures on the Phenomenology of Spirit*, ed. A. Bloom, trans. J.H. Nichols, Jr. New York/London: Basic Books.

Land, N. 1992. *The thirst for annihilation: Georges Bataille and virulent nihilism*. London: Routledge.

Libertson, J. 1982. *Proximity: Levinas, Blanchot, Bataille and Communication*. The Hague: Martinus Nijhoff.
Lucy, N. 2004. *A Derrida dictionary*. Oxford/Victoria: Blackwell.
———. 2000. Why does society describe itself as postmodern? In *Observing complexity: Systems theory and postmodernity*, ed. W. Rasch and C. Wolfe, 35–49. Minneapolis: University of Minnesota Press
Mauss, M. 1990. *The gift: The form and reason for exchange in archaic societies*. Trans. W.D. Halls. New York: Norton.
Morin, E. 2008. *On complexity*. Trans. S.M. Kelly. Cresskill: Hampton Press.
———. 1992. *Method: Towards a study of humankind*, vol. 1, *The Nature of Nature*. Trans. and intro. J.L.R. Bélanger. New York: Peter Lang.
———. 1973. *La Paradigme perdu: la nature humaine*. Paris: Seuil.
Müller-Lauter, W. 1971. *Nietzsche: Seine Philosophie der Gegensätze und die Gegensätze seiner Philosophie*. Berlin: Walter de Gruyter Verlag.
Nancy, J.-L. 1991. *The inoperative community*, ed. P. Connor, trans. P. Connor, L. Garbus, M. Holland, and S. Sawhney. Minneapolis: University of Minnesota Press.
———. 2003a. Concealed thinking. In *A finite thinking*, ed. S. Sparks, trans. J. Gilbert-Walsh, 31–47. Stanford: Stanford University Press.
———. 2003b. The unsacrificable. In *A finite thinking*, ed. S. Sparks, trans. R. Stamp and S. Sparks, 51–77. Stanford: Stanford University Press.
———. 2003c. A finite thinking. In *A finite thinking*, ed. S. Sparks, trans. E. Bullard, J. Derbyshire, and S. Sparks, 3–30. Stanford: Stanford University Press.
———. 2003d. Shattered love. In *A finite thinking*, ed. S. Sparks, trans. L. Garbus and S. Sawhney, 245–274. Stanford: Stanford University Press.
Nietzsche, F. 1967. *The will to power*, ed. W. Kaufmann, trans. W. Kaufmann and R.J. Hollingdale. London: Weidenfeld and Nicolson.
———. 1974. *The gay science*. Trans. W. Kaufmann. New York: Random House.
———. 1969. *Thus spoke Zarathustra*. Trans. R.J. Hollingdale. Harmondsworth: Penguin.
———. 1982. *Daybreak: Thoughts on the prejudices of morality*. Trans. R.J. Hollingdale, intro. M. Tanner. Cambridge: Cambridge University Press.
———. 1989a. *Beyond good and evil*. Trans. W. Kaufman. New York: Vintage Books.
———. 1989b. *Ecce homo*, ed. and trans. W. Kaufmann. New York: Vintage Books.
———. 1999a. *Sämtliche Werke. Kritische Studienausgabe in 15 Bänden*, vol. 10, ed. G. Colli and M. Montinari. Munich/Berlin: DTV/De Gruyter.
———. 1999b. *Sämtliche Werke. Kritische Studienausgabe in 15 Bänden*, vol. 11, ed. G. Colli and M. Montinari. Munich/Berlin: DTV/De Gruyter.
Poli, R. 2013. A note on the difference between complicated and complex social systems. *Cadmus* 2(1): 142–147.
Roodt, V. 2005. Amor fati, amor mundi: Nietzsche and Arendt on overcoming modernity. Doctoral dissertation, Stellenbosch University, Stellenbosch.
Shershow, S.C. 2001. Of sinking: Marxism and the "general" economy. *Critical Inquiry* 27(3): 468–492.
Spivak, G.C. 1994. Responsibility. *Boundary 2* 21(3): 19–64.
Stoker, W. 2012. Culture and transcendence: A typology. In *Looking beyond? Shifting views of transcendence in philosophy, theology, art, and politics*, ed. W. Stoker and W. van der Merwe, 5–29. Amsterdam: Rodopi.
Trahair, L. 2001. The comedy of philosophy: Bataille, Hegel and Derrida. *Angelaki: Journal of the Theoretical Humanities* 6(3): 155–169.
Van Tongeren, P. 2000. *Reinterpreting modern culture: An introduction to Friedrich Nietzsche's philosophy*. Indiana: Purdue University Press.
Weiss, A.S. 1986. Impossible sovereignty: Between "The Will to Power" and "The Will to Chance". In *Georges Bataille: writings on laughter, sacrifice, Nietzsche, Unknowing*. Trans. A. Michelson, 128–146. Massachusetts: The MIT Press.

Chapter 4
The Status and Ethics of Knowledge

[R]eal situations of communication do not depend only on the common code and repertory which is language; they also depend on another code, linked to ideology, which depends on a paradigmatology always implicit, always hidden, always present and always dominant.

- Edgar Morin (1977), La Méthode, v.1: La Nature de la nature

Abstract

The only complete description of a complex system is the system itself. However, since we cannot understand complexity in all its complexity, we are forced to model complex systems in order to render them meaningful. Models therefore reduce complexity, and are both an outcome of our mental constructions and empirical considerations. Complexity thinking necessitates an engagement with the nature, limits, and consequences of knowledge production, which raises both technical and ethical considerations.

In this chapter, this complexity-informed view of knowledge is further explored at the hand of philosophical positions that are sensitive to the normative dimension of knowledge creation. Specifically, Foucault's work on archaeology and genealogy is introduced in order to demonstrate how our knowledge practices are informed by largely unconscious epistemic frames, are construed on the basis of unreliable premises, and develop according to complex historical processes. The implications that Derrida's deconstructive philosophy hold for drawing attention to, and dealing productively with, the limited status of our models and the impossibility of the closure of meaning are also explored.

Both these philosophical positions reinforce the view that the epistemology of complexity cannot be construed as a value-free programme. The very real consequences that this position holds are explored at the hand of the question of the foreigner, who we seek to understand, but who—by definition—resists assimilation into our conceptual schemas.

M. Woermann, *Bridging Complexity and Post-Structuralism*,
DOI 10.1007/978-3-319-39047-5_4

4.1 Introduction

Questions concerning the manner in which knowledge is acquired and the status of our knowledge claims are central to philosophical reflection. Although today there is a higher level of sensitivity for the implications that complexity holds for our understanding of the world, the pervasive responses to the issues around knowledge reflect a deeply embedded subject-object dichotomy, whereby it is either believed that our understanding of the world is dependent upon our subjective and cultural experiences and psychological make-up, or that the world can be made rationally transparent and can yield objective and universal knowledge. In other words, most analytic models tend to separate the subject from the object (Cilliers 2000).

Cilliers however argues that from a complexity perspective the subject-object dichotomy is not helpful, since the subject is *constituted* within a network of knowledge; or otherwise stated, the subject is embedded in a context.[1] This means that we can neither isolate an objective context nor simply argue that the context is the product of subjective experiences. In this regard, Cilliers' views are well-aligned with Morin's (2008: 91), who defines his epistemological orientation as follows:

> Personally, I am a co-constructivist, which means that I think we construct our perception of the world, but with the help of the world which, as it were, lends us a hand.

The co-constructivist viewpoint is also supported by the fact of operational-closure, (which, to recall from Chap. 2, means that we access the world through our sensory apparatus), as well as by Kant's distinction between noumena and phenomena (whereby, as explained in Chap. 3, noumena or the things-in-themselves allow us to experience phenomena, but do not correspond with these phenomena). What the above means is that we cannot 'evade the status of knowledge as "translation" and "construction"' (91). The act of knowledge acquisition is not only facilitated by the translation of environmental stimulus, but also by the active reduction of complexity, through means of modelling. Following Cilliers (2001), I define models as including ideas, theories, paradigms, and systems of rules. Models are necessary because we are incapable of dealing with complexity in all its complexity. Models reduce complexity, but simultaneously create distortions of the system under study by framing a problem in a specific and limited way. Complexity thinking therefore implies an engagement with the nature and limits of models.

To say that we cannot accurately model a complex world also does not mean that all models are equally good or useful. Some models resonate better with our experiences of the world than others do, even though the gap between the world and our models of the world can never be effaced. Cilliers (138) argues that the reality of this gap should 'serve as a creative impulse that continually challenges us to transform our models, not as a reason to give up.' Viewed in this light, a model is more like a novel than a formula in that:

[1] This view of identity will be explored in more detail in the following chapter.

> like a novel, [a model] may resonate with nature, but it is not the "real" thing. Like a novel, a model may be convincing—it may "ring true" if it is consistent with our experiences of the natural [and the social] world. But just as we may wonder how much the characters in a novel are drawn from real life, and how much is artifice, we might ask the same of a model: how much is based on observation and measurement of accessible phenomena, how much is based on informed judgement, and how much is based on convenience? (Oreskes et al. 1994: 644).

Although not directly referring to the problem of complexity, Naomi Oreskes and her colleagues are sensitive to the fact that we do not have full access to the subjects of interest. Modelling therefore introduces questions of convenience and judgement, neither of which concern objective descriptions, but are rather normative issues. As previously argued, how we frame systems[2] (in other words, the boundaries that we draw around systems) is not only a function of the activity of the system itself, but is also a product of the description that we give to the system (Cilliers 2001). Otherwise stated, and as noted by von Foerster,[3] the observer is implicated in her observations. This insight resonates well with Luhmann's (2000) understanding of boundaries or frames, which, to recall from the previous chapter, he defines as 'the self-produced and reproduced difference between a system and its environment' (46). We do not frame objective reality, but our own observations. Conceptually speaking, there is no logical way out of this problem, or otherwise stated, there is no grand *Theory of Frames*.

Rosen (2005: 42) also remarks on this problem in his description of the notion of systemhood in science in which the environment is typically contrasted with the system. In this regard, he writes: '[t]he partition of ambience into system and environment, and even more, the imputation of that partition to the ambience itself as an inherent property thereof, is a basic though fateful step for science'. Although partitioning the world into systems and environments is a fateful step, it is also a necessary step, since we cannot know the world in all its complexity. In other words, boundaries are also *enabling*: we need to construct boundaries to demarcate systems from their environments, even though these boundaries introduce further complexities and uncertainties (since our descriptions are imperfect).[4] Therefore, in accordance with the co-constructivist viewpoint, boundaries are both physical properties of systems and mental constructions or idealisations. Conceding to both complexity and the necessity of modelling therefore commits us to weak reductionism, as opposed to strong reductionism. Weak reductionism means that one acknowledges the necessity of modelling, yet one does not harbour any illusions that models accurately and formally represent the system under study.

In this chapter, I shall explore philosophical positions that accord with the above understanding of knowledge (as the outcome of limited models). Specifically, I shall

---

[2] Ontologically-speaking one can forego the distinction between a system and its environment (in this regard, recall Nancy's view of radical immanence). However, from the perspective of epistemology, modelling is unavoidable as meaning is contingent on frames and models.

[3] See Sect. 2.2.2.

[4] See Cilliers (2001).

investigate Foucault's archaeological and genealogical methods and Derrida's deconstructive philosophy. In developing his methodology, Foucault engages critically with, and poses an alternative to, both hermeneutics and structuralism. These latter philosophical movements are generally viewed as the two most influential responses to the Kantian subject-object divide (Dreyfus and Rabinow 1983); and, as such, begin to address the shortcomings of the standard responses to issues of knowledge, as outlined in the opening paragraphs of this chapter. In seeking to overcome the weaknesses of both these responses, Foucault develops a highly original and influential response to the question concerning the nature of knowledge. His treatise constitutes a convincing philosophical exposition of the limited and exclusionary nature of knowledge, as well as of our inclination to overlook this fact.

Derrida also addresses the question of knowledge. Apart from sharing Foucault's interest in the structuralist project, he however does so along very different lines to Foucault. Derrida develops his deconstructive philosophy in order to deal productively with the necessarily incomplete nature of meaning (which, to recall,[5] he views as the consequence of the *aporetic* logic that defines meaning). Derrida employs deconstruction to expose the limited and exclusionary status of models that have become naturalised over time, and to upset the oppositional logic upon which these models rely. This is achieved by focusing on the play of *différance* that cannot be arrested and that prevents conceptual closure.

In order to illustrate the ethical implications that this view of knowledge holds, a personal anecdote, which illustrates the difficult problematic of the foreigner, will be presented. The various issues pertaining to our normal reactions to foreignness (as that which lies beyond or outside of our models, but nevertheless presents itself to our models) will be explored at the hand of Derrida's treatment of the question of the foreigner and of hospitality.

4.2 Foucault

4.2.1 Foucault's Archaeology and Genealogy

As mentioned above, Foucault's project engages with, but also seeks to transcend, the structuralist and hermeneutic approaches to meaning. This section opens with a brief description of structuralism, so as to better be able to identify both the structuralist elements in Foucault's work, and the differences between a structuralist and a Foucauldian account of meaning.

Louis Markos (2012) defines structuralism as:

> An interdisciplinary approach to all branches of human knowledge that rejects all ontological and epistemological sources of meaning in favour of an anti-metaphysical approach that posits that all humanistic pursuits are the product of deep structures that precede human consciousness.

[5] See Sect. 3.4.1.

Markos proceeds to unpack the definition as follows: structuralism is anti-metaphysical and anti-epistemological because the source of meaning is neither generated from above (Plato's ideal forms), nor from within (a view that gave rise to Descartes' and Kant's subject-object dichotomy). Rather, meaning is the product of the pervasive structures that undergird discourse and practice in general. Meaning is therefore generated from below, and is external to the agent. Markos explains that this implies that we are not the makers of our own destiny, for the reason that language determines thought—not the other way around. The structuralists thus shifted the source of meaning from a meaning-giving transcendental subject to objective and determinable structures (which are the product of basic elements, and the rules and laws by which they are combined). In so doing, the structuralists sought to re-establish the humanities on a more scientific base. Structuralism enjoyed immense popularity (especially in France) during the 1960s and its influence is visible in the twentieth century movement away from the humanities to the human sciences (Markos 2012).

According to the Foucauldian scholars, Hubert Dreyfus and Paul Rabinow (1983: 15), *The Birth of the Clinic* represents Foucault's 'extreme swing towards structuralism'. In this work, Foucault (1975: 198) attempts to show that 'the figures of knowledge and those of language... obey the same profound laws', i.e. the laws of structure. In this text, Foucault presents an archaeology of medicine, wherein he gives an archaeological account of the 'ineradicable chronological threshold' (195) between the Classical Age and the Modern Age of man. The ineradicable threshold signifies the seismic shift from one epoch to the next, whereby the constituting rules upon which our practices are based are radically transformed.

Before turning to the archaeological account, let us first consider the official explanation given for the shift from one epoch to another, namely that 'medicine finally broke away from fantasy and superstition and arrived at objective truth about the body and its diseases' (Dreyfus and Rabinow 1983: 12). From the vantage point of modern medicine (which is deemed objective and scientific), older medical accounts such as the following description provided by Foucault (1975: ix) seem incomprehensible and shocking:

> Towards the middle of the eighteenth century, Pomme treated and cured a hysteric by making her take "baths, ten or twelve hours a day, for ten whole months." At the end of this treatment for the desiccation of the nervous system and the heat that sustained it, Pomme saw "membranous tissue like pieces of damp parchment... peel away with some slight discomfort, and these were passed daily with the urine; the right ureter also peeled away and came out whole in the same way." The same thing occurred with the intestines, which at another stage "peeled off their internal tunics, which we saw emerge from the rectum. The oesophagus, the arterial trachea, and the tongue also peeled in due course; and the patient had rejected different pieces either by vomiting or by expectoration.

The notion that current medical accounts are more progressive and sophisticated than the account offered above is however not simply assumed by Foucault. By employing the archaeological method, Foucault subjects all discourse and knowledge—including medical knowledge—to 'the same sort of distanciation of truth and meaning which we naturally bring to the medical accounts and other theories of the Classical Age' (Dreyfus and Rabinow 1983: 13). In using this method, Foucault is able to uncover the rules of formation and the codes of knowledge that define our own Age, and thereby to deliver a critique of our smug assumption that medical science—or any other discipline—has advanced to an objective truth. In Foucault's (1975: 195) words:

> the new medical spirit… cannot be ascribed to an act of psychological and epistemological purification; it is nothing more than a syntactical reorganization of disease.

The above implies that the shift from the Classical to the Modern epoch concerns a change in the syntactic form of knowledge, rather than in semantic content. The archaeological method is used here to treat medical accounts as essentially meaningless objects in order to uncover the code that gives rise to these accounts. This allows us to see that that which remains incomprehensible from the perspective of our own epoch 'is not without its own systemic order' (Dreyfus and Rabinow 1983: 13).

Foucault's account comes very close to a structuralist account, in which meaning is viewed as the product of deep structures underlying discourse. However, Dreyfus and Rabinow also argue that, despite claiming in *The Birth of the Clinic* that archaeology could discover 'deep structure' (Foucault 1975: xix), Foucault was never truly a structuralist as he was more interested in '*historical*… conditions of possibility' (xix) rather than atemporal structures. Indeed, Foucault criticised the structuralist's reduction of human behaviour to rule-governed transformations of essentially meaningless entities (Dreyfus and Rabinow 1983). He also vehemently renounced the label of 'structuralist' accorded to him by—what he calls—'certain half-witted 'commentators'' (Foucault 2002a: xv).

In the foreword to the English edition of *The Order of Things: An Archaeology of the Human Sciences*—Foucault's next major work—he (Foucault, 2002a: xv) writes:

> There may well be certain similarities between the works of the structuralists and my own work. It would hardly behove me, of all people, to claim that my discourse is independent of conditions and rules of which I am very largely unaware, and which determine other work that is being done today. But it is only too easy to avoid the trouble of analysing such work by giving it an admittedly impressive-sounding, but inaccurate, label.

In this work, Foucault shifts his analysis from specific institutions (such as the clinic or the asylum) and concentrates instead on discourse practices as they pertain to the science of man (i.e. the human sciences). Foucault is particularly interested in the systems of self-understanding in Western thought, and his aim is to rediscover on what basis knowledge and theory become possible (Dreyfus and Rabinow 1983). To this end, he isolates epistemic systems, which he defines as 'the total set of relations that unite, at a given period, the discursive practices that give rise to epistemological figures, sciences, and possibly formalized systems' (Foucault 2002b: 211). He analyses these systems at the level of discursive realities. The epistemes that particularly interest Foucault in this work are the Renaissance, the Classical Age, and the Age of Man. Although it is beyond the scope of this work to provide a detailed analysis of this work, his own reflection on this project (as presented in the foreword to *The Order of Things*) is worth recounting as it sheds further light on the archaeological method.

In this reflection, Foucault (2002a) raises four points for our consideration. Firstly, he states that the hypothesis with which he is working is that the history of non-formal knowledge obeys the laws of a specific code of knowledge. This is despite the fact that non-formal knowledge (e.g. the science of living beings, languages, or economic facts) has been relegated to the margins of the history of the science of thought for the reason that these disciplines are believed to be too tinged with empirical thought and contingency.

Secondly, Foucault defines *The Order of Things* as a comparative, as opposed to a symptomatological, study. His claim is that the spirit of a century is not found in a particular body of knowledge, but instead in comparative readings of the different elements of non-formal knowledge. These comparative readings are aimed at relating non-formal knowledge to the philosophical discourse of the time. The emphasis is thus on the isomorphisms that emerge despite the extreme diversity of the objects under study. By abandoning disciplinary divisions, one is able to organise knowledge differently. Instead of describing the genesis of science (i.e. the original mode of formation), Foucault is interested in articulating the epistemological space of a specific period.

Thirdly, Foucault operates at the level of the archaeologist, as opposed to the historian of science. The historian focuses on scientific consciousness in order to try and restore that which has eluded this consciousness (i.e. the negative, unconscious underbelly of science that resists or disturbs science). Foucault (xi–xii), in contrast, is interested in revealing 'a *positive unconscious* of knowledge', which he defines as 'a level that eludes the consciousness of the scientist and yet is part of scientific discourse, instead of disputing its validity and seeking to diminish its scientific nature.' Foucault argues that that which the naturalists, economists, and grammarians of a certain epoch share are the rules of formation used to define, form, and build theories and concepts, even though the rules themselves are not formulated in their own right. These rules of formation thus determine the basis or archaeological systems common to a whole series of disciplines within a certain epoch.

Lastly, Foucault defines his work as 'an open site' (xii), but also argues that he disengages with three problems, namely the problems of change, causality, and the subject. As concerns change, Foucault argues that historical transformations are not the result of a single person, event, or collective spirit, but of a combination of corresponding transformations. As concerns causality, Foucault argues that he is wary of speculating about the specific reasons for specific changes, and has thus opted to bracket the problem of causes, and focus instead on describing the transformations themselves. As concerns the subject, Foucault concedes that the archaeological method takes the human out of the picture. He explores the human sciences neither from the viewpoint of meaning-giving individuals (phenomenology) nor from the perspective of formal structures (structuralism), but in terms of the rules that come into place in the very existence of discourse. These rules serve to validate certain discourses at certain times.

Ultimately, it was both the inability of the archaeological method to account for causes that give rise to discourse objects and practices in the human sciences, and the archaeologist's inability to account for her own involvement in discourse that led to a shift in Foucault's methodology.[6] Although Foucault never entirely abandons archaeology, he subordinates archaeology to genealogy in his later works. Whereas the object of study in archaeology is the human sciences as discourse-object; in genealogy, the human sciences are both the source and the outcome of a larger set of organised and organising practices (of which theory is but one of the components) (Dreyfus and Rabinow 1983). Whereas the archaeologist assumes the role of a neutral and disinterested observer; the genealogist is involved in, and co-produces, the practices that she is studying. Aside from introducing genealogy in order to address the shortcoming of archaeology, Foucault also amended his methodology so as to better deal with the question of power. Issues concerning power constitute the focus of this later work. Foucault contends that power could not be properly studied in theoretical form, or as the outcome of a paradigm. Rather, power is enmeshed in our organised and organising practices that are themselves the outcome of complex genealogical processes (Foucault 1990).

Before describing his genealogical method in more detail, it is useful to investigate a second early influence on Foucault, namely hermeneutics. The reason for this is that the genealogical method should, in part, be understood as a reaction to hermeneutics (in which the source of meaning shifts from a transcendental phenomenological subject to cultural and historical practices).

Hermeneutics is an important theme for Heidegger, who, in *Being and Time*, posits that our phenomenological experience of being-in-the-world is informed by our historical and cultural practices that form a backdrop to our lived experiences. These practices—which Heidegger (1962) terms *interpretations*—are never made entirely explicit to the subject. To understand these everyday practices, further interpretation is needed. Heidegger terms the interpretation of the interpretations of our everyday practices *hermeneutics*. In *Being and Time,* Heidegger distinguishes between two types of interpretations. The first type concerns an interpretation of being in its everydayness (as informed by our partial and distorted understanding of our practices). It constitutes a 'motivated masking of the truth' (xxii) in that we are so caught-up in beingness and in the comings and goings of our daily lives that the true nature of Being withdraws. The second type acts as a corrective to the first type. Through the work of this second type of interpretation—understood as a counter-movement to Being's concealment—a deeper primordial truth concerning the true nature of Being comes to the fore. In Heidegger's (359) words:

> Dasein's *kind of Being… demands* that any ontological interpretation which sets itself the goal of exhibiting the phenomena in their primordiality, *should capture the Being of this entity, in spite of this entity's own tendency to cover things up.*

Heidegger defines the Being of beings (or the true entity of being) as the essential groundlessness of being. For Heidegger, hermeneutics thus concerns repetitive—yet progressive—acts of interpretation, designed to lead us ever closer to the heart of being, which is nothing other than Being itself. Foucault however is wary of the

[6] See the next section for a more detailed discussion of these issues.

hermeneutic task, which he views as regressive. Foucault understands hermeneutics as the attempt to recover truth through means of interpretation or *commentary*. He defines commentary as 'the re-apprehension through the manifest meaning of discourse of another meaning at once secondary and primary' (Foucault 2002a: 407). Although Foucault (1973) comes dangerously close to identifying madness as the deep secret of experience in the conclusion to *Madness and Civilization*, he ultimately rejects the hermeneutic approach. The reason for this is that he does not believe that our practices have an implicit and deeper meaning, and he thus views commentary as an endless and fruitless exercise.

Foucault's genealogical method, which characterises his later works, and which is largely influenced by Nietzsche's understanding of genealogy, is diametrically opposed to the idea of a hidden meaning. In this regard, Foucault (1967: 187) writes that:

> Whereas the interpreter is obliged to go to the depth of things, like an excavator, the moment of interpretation [genealogy] is like an overview, which allows depth to be laid out in front of him in a more and more profound visibility; depth is resituated as an absolutely superficial secret.

For the genealogist, there can be no uncovering of hidden meaning, precisely because there is no primary base to interpret. Interpretation is a never-ending task, leading to a proliferation of interpretations (Dreyfus and Rabinow 1983). Given this view, Foucault (1984: 87–88) writes that '[n]othing in man—not even his body—is sufficiently stable to serve as the basis for self-recognition or for understanding other men'. Foucault defines the body as 'the inscribed surface of events' (83), in which diverse elements find their sudden expression, and violently engage in order to efface one another. The body is therefore an emergence and a site of conflict. It is also representative of larger historical events that are themselves the outcome of disparate and competing forces. *Place* should thus be understood as an emergence or the entry of forces that erupt 'from the wings to center stage, each in its youthful strength' (84).

Given the fact that there is no essence to our existence, the genealogist turns her attention to seeking out the discontinuities rather than the continuities, recurrences and play rather than progress and seriousness, and the surface of events rather than the depth of meaning (Dreyfus and Rabinow 1983). Foucault (1984: 81) writes that the aim of genealogy is:

> to follow the complex course of descent [in order] to maintain passing events in their proper dispersion; it is to identify the accidents, the minute deviations—or conversely, the complete reversals—the errors, the false appraisals, and the faulty calculations that gave birth to those things that continue to exist and have value for us; it is to discover that truth or being does not lie at the root of what we know and what we are, but the exteriority of accidents (81).

Foucault (76–77) further states that '[g]enealogy… requires patience and a knowledge of details, it depends on a vast accumulation of source material… In short, genealogy demands relentless erudition.' This task of relentless erudition is what Foucault calls writing effective history (*wirkliche Historie*). Foucault contrasts effective history with traditional history, and summarises the differences between these positions as follows: firstly, whereas historians traditionally attempt to construct a comprehensive view of the past, the genealogist places 'within a process of development everything considered immortal to man' (87) in order to highlight the contingency of current practices and events. Secondly, the genealogist deals with events in terms of their unique attributes in order to dispel the myth that the course of history is shaped by 'profound intentions and immutable necessities' (89), as traditional historical accounts would have us believe. Thirdly, effective history constitutes a study of nearby things, as opposed to distant events. Unlike traditional historical accounts that paint a picture of the past in the present, effective history concerns the writing of the history of the present. And, lastly, in contrast to historians that 'take unusual pains to erase the elements in their work which reveal their grounding in a particular time and place' (90); the genealogist, writing effective history, affirms knowledge as perspectival.

Foucault's interest in genealogy is closely tied to his interest in the interrelated problems of power, knowledge, and the body. In this regard, effective history is the means by which Foucault traces a particular ritual of power or political technology of the body that manifests in the present back into the past in order to show how a specific practice took on different meanings at different times (Dreyfus and Rabinow 1983). One such a ritual of power that Foucault explores at length is that of sexual taboos. In the opening pages of *The History of Sexuality (vol. 1)*, Foucault (1990) argues that modern day sexual norms and taboos are the outcome of a history of sexuality, through which sex itself has become the object of scientific discourse and inquiry. In this regard, Foucault (12–13) hypothesises that:

> the "putting into discourse of sex," far from undergoing a process of restriction, on the contrary has been subjected to a mechanism of increasing incitement; that the techniques of power exercised over sex have not obeyed a principle of rigorous selection, but rather one of dissemination and implantation of polymorphous sexualities; and that the will to knowledge has not come to a halt in the face of a taboo that must not be lifted, but has persisted in constituting—despite many mistakes, of course—a science of sexuality.

Foucault states that the goal of *The History of Sexuality* is to explore these movements at the hand of 'certain historical facts that serve as guidelines for research' (13). In other words, Foucault sets out to write the effective history of sex.

As previously stated, Foucault's genealogy was influenced by Nietzsche. Foucault specifically endorses Nietzsche's challenge to 'the inviolable identity of… origin' (Foucault 1984: 79), his understanding of history as 'the violent or surreptitious appropriation of a system of rules' (86), and his understanding of place or non-place as an '[e]mergence produced through a particular stage of forces' (83). However, whereas Nietzsche (1989: 152) wanted to expose truth as 'errors, blindness, [and] lies' wielded by individual actors, Foucault is concerned with 'how both scientific objectivity and subjective intentions emerge together in a space set up not by individuals but by social practices' (Dreyfus and Rabinow 1983: 108); as well as with how this space gives rise to specific practices of domination, power, and knowledge.

4.2.2 *Two Levels of Modelling*

When it comes to the study of human beings, Foucault is specifically interested in how we understand ourselves as intentional subjects, whilst simultaneously construing ourselves as objects (and therefore as the subject of knowledge). Already, this divided human, defined as both subject and object, betrays the difficulty mentioned in the introduction to this chapter, namely that of keeping the subject and object apart. Furthermore (and as also argued in the introduction), a consequence of foregoing the strict delineation between subject and object is that we are compelled to accept a view of the subject as constituted within a network of knowledge. For Foucault (1979), this network of knowledge operates in conjunction with power. The conjunction of power and knowledge should however not be understood as:

> the activity of the subject of knowledge that produces a corpus of knowledge, useful or resistant to power, but power/knowledge, the processes and struggles that traverse it and of which it is made up, that determines the forms and possible domains of knowledge (Foucault 1979: 28).

The network of *power/knowledge* is thus both the source and the outcome of our organised and organising practices. This view has its roots in Foucault's earlier work. Although the archaeological method does not account directly for power/knowledge, it does allow us to identify and analyse the network of knowledge—what Foucault (2002a: x) calls 'a certain code of knowledge'—of a given epoch. Foucault's epistemes therefore constitute models of various epochs, which—as previously defined—prescribe 'the total set of relations that unite, at a given period, the discursive practices that give rise to epistemological figures, sciences, and possibly formalized systems' (Foucault 2002b: 211). The episteme therefore constitutes the overarching rules and conditions (i.e. the frame) that informs our diverse disciplinary practices. These rules and conditions are interpreted in each of the disciplinary practices in order to develop discipline-specific models. Foucault thus distinguishes between two levels of modelling, i.e. the epistemic models and the models of our formal and non-formal knowledge. The latter models are informed by, and therefore dependent on, a given epistemic model. However, we can only access this epistemic model through undertaking an archaeological and genealogical study of our knowledge models.

This dual understanding of knowledge creation introduces a dimension of modelling that is not accounted for in the introductory discussion. If it is indeed the case that our knowledge models are informed by a broad episteme, then these models are not only the product of convenience, judgement, and the resources at our disposal, but also of the rules of formation that define the Zeitgeist of a specific episteme. The subject (modeller) remains largely unaware of these rules of formation, which constitute, what Foucault (2002a: xi–xii) calls, 'the positive *unconscious* of knowledge'.

This adds yet another level of complexity, since the epistemic frame that guides us (and that is dynamically and reciprocally constituted by, and constitutes, our practices) is largely unknown to us. The epistemic frame is therefore not accounted for in our practices, since even though our practices participate in an episteme, they cannot embody the episteme. In a certain sense, the boundary or frame of an episteme is inaccessible *as* boundary or frame—it can only be known through our practices, which nevertheless do not constitute the frame.

This paradox of framing is most clearly explained in Derrida's (1980) work on genre, in which he argues that even though a text always belongs to a genre, the frame that marks the genre does not itself belong. Derrida uses the example of a novel, which may or may not bear the subtitle, 'A Novel'. This designation of novel (which is either explicit or implicit) is not, in itself, novelistic; although it speaks of the corpus, from which it is neither included, nor excluded. The genre or limit thus frames a work, whilst being itself excluded from the work. In 'The parergon', Derrida (1979: 39) writes, '[t]here is framing, but the frame does not exist.'

This understanding of epistemic frames exacerbates Luhmann's position on the impossibility of a grand *Theory of Frames*. Constructing a Theory of Frames is not only impossible because of the regression problem (which is Luhmann's point[7]), but also because an epistemic frame does not conceptually exist as frame.[8] Rather, the frame occupies a type of (non)place, and it is the product of a collection of transformations that cannot be causally traced. Although not explicitly stated, Foucault understood that the frame itself was contingent on complexity and could therefore neither be reduced to atemporal structures that could be scientifically studied (as the structuralists would have it), nor be viewed as the product of a number of identified causes that could be attributed to the actions of specific agents.

Not only did Foucault disregard individual agents in his study of historical epochs, but himself as well. In other words, Foucault the archaeologist is not implicated in his archaeologies, which—as previously stated—constitutes one of the weaknesses of this method. Indeed, Roland Bélanger (1992), the English translator of Morin's *Method*, argues that, when compared to the works of the two great philosophical theorists of knowledge of our time, namely Foucault and Hans George Gadamar, Morin's project is clearly pitted 'against Foucauldian "archaeology" and for Gadamerian hermeneutics' (xx) for the reason that Foucault is unable to account for the role of the observer in his early work. Foucault's refusal to concede to his own role in his philosophical project may have something to do with an implicit awareness of the paradox of framing. The fact that the genre does not allow itself to be pulled into the contingency and historicity of the work that it ultimately describes is evidently a difficulty with which Foucault grappled. He was only able to overcome this difficulty by shifting his focus from archaeology to genealogy, which allowed him to account for both causality (albeit a complex causality), and the subject's involvement in meaning-formation.

---

[7] See Sect. 3.2.3.

[8] This may also well be the reason why we tend to naturalise our positions, i.e. we tend to see our current knowledge practices as neutral and objective when compared to the knowledge practices of other epochs, because we are blind to the frames that inform our current day epoch.

I read Foucault's genealogy as an attempt to further denaturalise our knowledge practices in order to account for the workings of power. Foucault is at pains to demonstrate that our practices neither correspond with a transcendental truth nor veil a deeper, hidden truth. Rather, meaning is dynamic, placing 'within a process of development everything considered immortal to man' (Foucault 1984: 87). There is no origin of meaning, merely a proliferation of meanings constituted by our practices. Nevertheless, Foucault tries to detangle these meanings in order to account for the development and value of certain current day practices, specifically as concerns power and the workings of power on the site of the body. To this end, Foucault attempts to trace the meaning of specific practices (such as sex) across epochs in order to account for the status and meaning of these practices in the present.

In tracing the meaning of these practices, Foucault introduces causality into his theory. On a Foucauldian account, causality is not understood as 'an unbroken continuity' (81). Rather, and as previously cited, causality is reinscribed in terms of a 'complex cause of descent' characterised by 'accidents', 'minute deviations' and 'complete reversals' (81). This complex cause of descent furthermore gives rise to our practices, which are themselves the product of 'errors', 'false appraisals', and 'faulty calculations' (81). However, despite the dubious origins and processes by which the present 'truths' of our practices come into being, Foucault nevertheless prizes these genealogical investigations since they shed a measure of light on 'those things that continue to exist and have value for us' (81). In this regard, Foucault's work resonates well with Morin's (1992: 361) insight that 'the progress of knowledge is at the same time the progress of ignorance', in that knowledge is the product of organisation, but also of noise and redundancy.

In summary, Foucault's philosophy holds important insights for us, specifically with regard to how our knowledge practices are informed by largely unconscious epistemic frames, are construed on the basis of unreliable premises, and develop according to a complex process. This view of knowledge formation raises normative questions regarding the status of our practices. Given this Foucauldian account of meaning, our practices cannot be construed as scientifically rigorous, neutral, or progressive. For Foucault, the central consequence of our practices (which he analyses primarily in terms of disciplinary technologies, which constitute the means for creating docile bodies) is that they give rise to bio-power. Bio-power concerns the management of populations via the instruments of the state, including 'the family and the army, schools and the police, individual medicine and the administration of collective bodies' (Foucault 1990: 141). Foucault (143) writes that 'one would have to speak of *bio-power* to designate what brought life and its mechanisms into the realm of explicit calculations and made knowledge-power an agent of transformation of human life.' Bio-power is a central theme in Foucault's later work. However, at this juncture I part ways with Foucault in order to turn to the work of Derrida, who also offers an account of non-naturalised meaning, as well as insight into how intervention is possible.

4.3 Derrida

4.3.1 Deconstruction and Intervention

It was argued in the introduction to this chapter that our inability to engage with complexity as such is due to the fact we have no direct access to reality. The subject cannot be separated from the object, and nor can the observer be separated from her observation. However, and following from Chap. 3, there is a second (related) reason for why our models are not categorical, objective, or universal, namely that meaning as such cannot be exhausted; or, otherwise put, meaning cannot correspond with the thing-in-itself. The excesses of the general economy always intervene to usurp any fixed status accorded to meaning.

To recall,[9] Derrida radicalises Kant in arguing that concepts not only do not correspond with noumena because of our inability to access the thing-in-itself, but also because of the aneconomic force of *différance*. The aneconomic force signifies the irreparable loss of meaning that disrupts every economy, that destroys the internal integrity of the thing-in-itself or the possibility of saturated meaning, and that threatens the total ruination of meaning. Derrida's engagement with this problematic comes most keenly to the fore in his work on the *aporetic* logic that guides all concepts. This *aporetic* logic boils down to the fact that 'every foundation established will already have been… ruined, from the instant of its institution' (Hurst 2004: 249). The reason for this ruination is the suspension of the aneconomic force of *différance*. This suspension is, in part, a consequence of modelling by virtue of which we arrest meaning. In this section, Derrida's understanding of the necessarily limited nature of conceptual models will be investigated, as will the implications that this view holds for intervening in these models.

For Derrida, meaning is necessarily interpreted and ordered. In other words—and as previously argued—reality as such cannot be understood, except through modelling or framing. The model or frame not only allows for the creation of meaning, but also constitutes our conceptions of interiority and exteriority. In this regard, Derrida's (1979) discussion on *parergonality* or framing that follows from his deconstruction of Kant's theory of the aesthetic is of particular significance. Unlike Kant, who tries to resolve the problem of the frame by describing it as a special *parergon* (i.e. a hybrid of outside and inside), Derrida shows that in separating the work of art (the *ergon*) from its environment, the *parergon* also *constitutes* the work of art as something that is differentiated from its environment.

Aside from informing our conceptions of interiority and exteriority (or of what counts and what does not count from the perspective of the system), the process of modelling also determines the relative weight of the variables within the system itself. Models or conceptual schemas are hierarchically ordered, which supports the

[9] See Sect. 3.4.1.

insight that complex systems are structured as opposed to flat.[10] This also means that within any conceptual hierarchy, certain terms are necessarily privileged over others. Although from a theoretical perspective this description is neutral; when considered in terms of the anthropo-social domain, the description holds ethical and political implications. The reasons for this are as follows:

Firstly, and as concerns the interpreted nature of meaning, it is ethically important to reflect on the stakeholders and interests that count, or do not count, from the perspective of our limited models. No single model can accommodate the interests of everyone, which means that no foundation can live up to the call and obligation that justice demands. Derrida (1999a: 79) remarks on this point in the following passage, in which he refers to the nature of the text, which can be read as a remark on the nature of context (as will be explained in a subsequent paragraph):

> I would say that the text is complicated, there are many meanings struggling with one another, there are tensions, there are overdeterminations, there are equivocations; but this doesn't mean that there is indeterminacy. On the contrary, there is too much determinacy. That is the problem.

Secondly, the fact that certain terms are necessarily privileged over others within our models means that the privileged term (what Derrida calls the logos) operates at the expense of the marginal or secondary term, which serves to secure the status of the logos as logos. The status quo is thus always the product of those whose interests are served by current conceptual models, and those whose interests are suppressed or assimilated by the powerful. Therefore, even though Derrida (2002a) recognises the fact that we cannot do without hierarchies—in this regard he writes that '... I do not think that there are nonhierarchical structures. I do not think they exist' (21)—he is nevertheless wary of authority and hierarchy. In this regard, he writes that:

> I have, it seems, a quasi-aesthetic aversion to authority and hierarchy... The aesthetic aversion has to do more with the fact that, most often, the most common forms of authority and hierarchy, of power and hegemony, have something in them which is vulgar, insufficiently refined, or insufficiently differentiated: thus my aversion to authority, in this case, is also an aversion to what is still too homogenous, insufficiently refined or differentiated, or else egalitarian (20).

The limited nature of our models, the domination of one term over another, and the assimilation of differences within these models cry out for intervention. It is in response to these issues that Derrida develops his deconstructive philosophy, which entails the simultaneous reversal and displacement of conceptual hierarchies.

Derrida (1988a: 4) is loath to define deconstruction in terms of positive content for the reason that 'all the predicates... which seem at one moment to lend themselves to this definition or to that translation, are also deconstructed or deconstructible, directly or otherwise'. Nevertheless, I will venture a couple of remarks on the manner by which a deconstruction proceeds. As mentioned above, all conceptual hierarchies contain a logos and a secondary term(s). In the hierarchy of patriarchy,

---

[10] See Sect. 2.3.

for example, men occupy the position of the logos, whereas women and children occupy the secondary position (they mark the fall, so to speak). Animals, on the other hand, fall wholly outside of the hierarchy, as demonstrated by the practice of flesh-eating.[11] In order to usurp this hierarchy, a conceptual reversal must take place whereby it is shown that the logos is dependent on the support of the terms occupying the secondary position. In terms of the patriarchal hierarchy, the common expression 'Behind every great man is a woman' comes to mind. If men are dependent on women for their status, then their superior position can no longer be rigorously defended, which introduces instability into the hierarchy, as now it is plausible that either women or men can occupy the commanding position. However, a deconstruction cannot stop at a reversal for the reason that a reversal does not guarantee a conceptual displacement. (As an example, consider the case of a failed revolution whereby one party is merely overthrown by another, without bringing about a true political transformation). In order for a deconstruction to thus be successful, both a reversal and displacement of meaning must take place simultaneously. The goal of a displacement is to reframe the terms of conceptual systems so that they are no longer defined in opposition to one another, but are rather understood differentially, whereby there is a consequent exchange of meaning between the terms (meaning is deferred from one term to the next). In order to explain the displacement, Derrida often makes use of an original arche-concept, which represents the properties that the terms in the hierarchy share. In our example, arche-woman would signify this original type. Both women and men would serve as sub-categories of the arche-woman in that they share the definitive traits defining the arche-woman.

Deconstruction is therefore a means by which one can intervene in these hierarchies or conceptual models, not with the goal of destroying these models, but in order to denaturalise and destabilise them and to allow qualitative differences to emerge. In Derrida's (2002a: 21) words, intervention or deconstruction is:

> not a way of nullifying the hierarchy but a way of destabilizing the given hierarchies and codes and of answering to the expectations, to the desire and the motivation of people who are oppressed... But in any case, I am not an enemy of hierarchy in general and of preference nor even of authority. I am simply impatient before the given, stabilized, installed, that is to say, vulgar and dormant forms of hierarchy.

The difficulty with deconstruction is however that we have no tools or resources with which to work outside of our existing conceptual structures. As Derrida (1976a: 163) explains, the reason for this is that '[t]here is nothing outside of the text.' This is not a constructivist claim; rather it means that 'there is nothing outside context' (Derrida 1988b: 136). There is no Archimedean point from which to undertake and evaluate our interventions. Given this state of affairs, Derrida cautions that every deconstruction should be preceded by a careful and close reading of context in order to ensure that one's interventions do not result in doing or saying 'just anything at all' (Derrida 1976a: 158). Derrida explicitly renounces relativ-

[11] See Sects. 5.3.1 and 5.3.2.

ism in stating that '...I have never accepted saying, or being encouraged to say, just anything at all' (158). In support of the necessity of understanding context, Derrida (1999a: 67) writes:

> I would assume that political, ethical and juridical responsibility requires a task of infinite close reading. I believe this to be the condition of political responsibility: politicians should read. Now to read does not mean to spend nights in the library; to read events, to analyse the situation, to criticize the media, to listen to the rhetoric of demagogues, that's close reading, and it is required more today than ever. So I would urge politicians and citizens to practice close reading in this new sense, and not simply to stay in the library.

It is therefore only on the basis of a close reading that responsible intervention can take place. Such a close reading necessarily requires vigilant and critical reflection. Yet, intervention also implies action. As such, the moment of reflection must also necessarily give way to a moment of action.[12] In 'The conflict of faculties', Derrida (1982 in Culler 1983: 156) affirms this in stating that:

> Deconstruction is also, at the very least, a way of taking a position, in its work of analysis, concerning the political and institutional structures that make possible and govern our practices, our competences, our performances. Precisely because it is never concerned only with signified content, deconstruction should not be separable from this politico-institutional problematic and should seek a new investigation of responsibility, an investigation which questions the codes inherited from ethics and politics.

In terms of analysing the political and institutional structures that govern our practices, deconstruction operates on two fronts simultaneously: it offers a critique of social and political institutions, and a critique of philosophical oppositions that enable these institutions. In this regard, Derrida acknowledges that an insurmountable gap exists between these two levels of intervention, and yet he argues that we should try to reduce this gap. In his words:

> we must take into account certain gaps and try to reduce them even if, for essential reasons, it is impossible to efface them: for example, between the discourses and practices of this immediately political deconstruction and a deconstruction with a theoretical or philosophical aspect. These gaps are sometimes so great that they dissimulate the relays or render them, for many, unrecognizable (Derrida 1976b: 113).

Whereas immediate political interventions are likely to lead to more conservative outcomes (because the effects of such interventions are, to an extent, calculable); the deconstruction of philosophical paradigms, on the other hand, may lead to unexpected consequences. This is because we cannot imagine ourselves 'renouncing all consciousness, all presence, all ethics of language: and yet this is precisely what must be deconstructed' (Derrida 2002a: 16). In other words, we cannot imagine giving ourselves wholly over to the general economy, even if we recognise the fact that our restricted, codified systems destroys the aneconomic force of the general economy. Due to the threat that deconstruction holds, Derrida argues that we tend to avert our eyes 'when faced by the as yet unnameable which is proclaiming itself... in the form-

---

[12] See Sect. 3.4.2.2.

less, mute, infant, and terrifying form of monstrosity' (Derrida 1978: 293). If we turn away from our fears, we run the risk of excusing 'a conservative institutional complacency' (Culler 1983: 159); yet, if we completely give into the need to safeguard the space of the general economy, we run the risk of sacrificing ourselves to 'the very menacing character of deconstruction' (Derrida 2002a: 16). It is due to this inevitable *aporia* that Derrida argues that we have a duty to try to reduce—but not to efface—the gap between these two levels of meaning, which also means that we should occupy the space of deconstruction or negotiation, understood as the place of threat:

> One must [il faut] with vigilance venture as far as possible into what appears threatening and at the same time maintain a minimum of security—and also an internal security not to be carried away by this threat (16–17).

The one horn of the *aporia* identified in this citation tempers the radical nature of deconstruction by preventing deconstruction from being equated with destruction. However, deconstruction also loses some of it radicality in the moment when critique translates into practice. The gap or delay mentioned above can therefore also be described as the gap between theory and application. Deconstruction is only meaningful in its application; but in applying, one becomes inextricably involved in the opposition or system that one wishes to criticise. In Derrida's (26) words: 'the position is not the affirmation, the very positivity of the institution will threaten, corrupt, cover over the affirmation'. To lose one's radical position also means to lose the critical disposition that feeds deconstruction. To prevent this outcome it is important not to treat the two horns of the *aporia* as two separate activities, but to practise them simultaneously. This entails that we should try to account for both the economic and aneconomic dimensions of meaning, by remaining cognisant of the fact that each foundation is ruined in the instant of its instantiation. In practical terms, this boils down to a refusal to separate politics from the discourse (or general text) in which it operates. As such, deconstruction politicises and problematises institutions to the extent that it draws attention to the connections between theory and practice, and thereby initiates a critical questioning of the code that is inherited from politics and ethics and that undergirds our institutions (Culler 1983).

The gap between practice and theory, or institutions and philosophy, also problematises the outcome of deconstruction to the extent that it is not clear what the outcome should be. For example, in seeking to dismantle patriarchy, feminists can try to deconstruct logocentric systems by challenging the opposition between males and females, and thereby minimise sexual differentiation; or, they can exalt in sexual differentiation, by stressing the opposition between males and females, and by celebrating the feminine (Culler 1983). The latter strategy involves demonstrating the female's independence from the conceptual apparatus sustaining the patriarchal hierarchy and 'male' modes of thought and behaviour. Again, there is no easy synthesis between these two outcomes. Yet, the necessity of working with the gap, and on two fronts simultaneously, remains.

4.3.2 Modelling and the Moment of the Decision

From the above, it is clear that the force of deconstruction lies in intervention. This analysis serves to debunk the opinions of critics, like Richard Rorty (1978), who claim that Derrida is merely interested in playing with words. To the contrary, deconstruction is ethical and political through-and-through, and the ethical and political consequences of deconstruction derive from Derrida's understanding of how meaning is formed in an immanent world. To summarise: meaning is the product of conceptual hierarchies, which are necessarily partial and exclusive, and which therefore require intervention. The goal of intervention is to denaturalise these hierarchies, by challenging the conceptual order, and by drawing attention to the qualitative differences and the play of *différance* that the naturalisation of such hierarchies suspends. Intervention is however not without challenges, and these challenges are directly derived from the manner in which meaning is modelled. Like Foucault, Derrida acknowledges two levels of modelling: one concerns the conceptual, philosophical level (which loosely corresponds to, but is also much broader than, Foucault's epistemic level), and the other concerns the institutional level (Foucault's level of knowledge models).

Foucault is interested in determining how epistemic rules of formation (of which we are largely unaware) give rise to knowledge practices (and vice versa), as well as in the consequences that these practices hold for our understanding of ourselves as both subjects and objects. Derrida, in contrast, explores the possibility of critically intervening in the philosophical preconditions upon which our knowledge practices and institutions are based, even though he acknowledges that it is impossible to renounce 'all consciousness, all presence, all ethics of language' (Derrida 2002a: 16). For Foucault, intervention is possible, but only at the level of individual practices. Indeed, in his later works, Foucault advocates an aesthetics of self-formation whereby the self exercises power over the self in a bid to resist the practices of subjugation that Foucault identifies as a consequence of modern bio-power. In problematising the relation between theory and practice, Derrida however seeks to intervene on both these levels, even if such interventions are governed by an intractable *aporia* that both unites and separates theory and practice. What Derrida tries to demonstrate through his work, is the necessity and the difficulty of intervention, which also translates as the necessity of, and difficulty inherent to, decision-making.

Derrida (31) states that intervention always implies a gamble, because 'one is never sure of the right time, there is always a risk.' To understand what Derrida means by this, it is important to follow him in his treatment of decision-making. Derrida distinguishes between a calculation (which is essential programmable) and a decision (which is non-programmable and incalculable). Since the decision cannot be programmed in advance, our decisions can lead to negative consequences that we were unable to foresee. The decision is thus governed by an inherent undecidability as to which course of action or intervention to pursue. Yet, this undecidability, and the risk that it carries, does not excuse us from making decisions;

rather, it *initiates* the decision as a responsible decision. In this vein, Derrida (1988b: 116) writes:

> A decision can only come into being in a space that exceeds the calculable programme that would destroy all responsibility by transforming it into a programmable effect of determinate causes. There can be no moral or political responsibility without this trial and passage by way of the undecidable. Even if a decision seems to take only a second and not be protected by any deliberation, it is structured by this *experience and experiment of the undecidable*. If I insist on this point from now on, it is, I repeat, because this discussion is, will be, and ought to be at bottom an ethical-political one.

I have quoted this passage at length,[13] because it holds important implications for our understanding of the ethics of modelling. Ethics is not merely a by-product of modelling, i.e. the argument that because models are partial, we have to take responsibility for the consequences of our models. Rather, and more radically, ethics is inaugurated in the moment of modelling, i.e. in the moment that we choose or decide to frame meaning in a specific way. Although the difference between these two points is subtle, it is important, as it defines ethics or responsibility as *a priori* part of the decision, rather than an *a posteriori* consequence of the decision.

4.4 Critical Complexity

This last insight also has consequences for how we approach complex phenomena. Cilliers (2010) argues that our approach should be governed by a self-critical rationality, which he defines as 'a rationality that makes no claim for objectivity, or for any special status for the grounds from which the claim is made' (14). As previously stated,[14] Cilliers views a self-critical rationality as central to complexity thinking. For this reason, he refers to his approach to complexity as *critical complexity* (Cilliers 2016). Although Morin (2007) frames his approach in terms of *general complexity*, he too is sensitive to the implications that an acknowledgement of complexity holds for our understanding of phenomena. This is clear in his juxtaposition of Cartesian doubt with the doubt generated from an awareness of complexity. In this regard, he writes:

> The Cartesian doubt was sure of itself. Our doubt doubts of itself; it discovers the impossibility of starting from scratch, since the logical, linguistic, cultural conditions of thought are inescapably prejudging. And this doubt, which cannot be absolute, can no longer be absolutely resolved (Morin 1992: 10).

Although Cilliers did not have the opportunity to fully systematise his understanding of critical complexity prior to his untimely death, an early article, titled 'Complexity, deconstruction and relativism' (2005), provides some additional

---

[13] Reference is again made to this passage and to Derrida's views on decision-making in Sects. 6.1 and 6.2, for the reason that decision-making is central to Derrida's views on ethical praxis.

[14] See Sect. 1.2.

insight into the implications that the critical stance holds for the manner in which we should approach complex phenomena.

Our traditional understanding of ourselves, the world, and others is typified by a certain assertiveness. This attitude finds its expression in the dictum *scientia potential est* (which is translated as knowledge is power, and which is attributed to Francis Bacon). Assertive positions favour a view of the world that is orderable, explainable, and controllable. These positions, and the concomitant worldviews that they give rise to, are premised on the failure to recognise complexity as they are based on a restricted approach to knowledge generation.[15] Cilliers (2005) argues that this failure constitutes not only a technical—but also an ethical—failure. This is because denying the exclusionary and partial nature of our models means that we negate not only issues of judgement and convenience or the epistemic and philosophical assumptions that inform our models, but also the fact that ultimately these models are based on decisions that are not programmable. Ethics is always already inherent to our practices, and Cilliers argues that to forget or to deny this leads to a negative ethics. Consequently, '[w]e must' as Morin (1992: 10) asserts, 'start by extinguishing false clarity.'

In contrast to assertive positions, critical positions that are sensitive to complexity offer a provisional stance to knowledge. These positions are often under pressure from more assertive positions, which claim to provide definitive answers to the world's problems. In defence of critical positions, Cilliers (2005) argues that acknowledging complexity and being modest about the scope of our knowledge claims does not commit us to either relativism or vagueness, for the reason that all positions or models are not equally acceptable. This means that it is possible to judge, evaluate, and possibly revise a decision once it has been made, even if the moment of the decision is characterised by undecidability. If this were not the case, intervention and transformation would be impossible. Therefore, conceding to the inevitable limits of our claims does not imply that there are limits to the intelligibility of our claims. In Cilliers (263) words: 'We can make strong claims, but since they are limited we have to be modest about them.' Peter Allen (2000) also supports the view that provisional models need not be viewed as relativist or vague. Indeed, he states that '[a] representation or model with no assumptions whatsoever is clearly simply subjective reality…. In this way, we could say that it does not therefore fall within the science of complexity, since it does not concern systemic knowledge' (93). However, instead of viewing models as absolute representations of reality, we must be aware of, and 'apply our "complexity reduction" assumptions honestly' (94) and take note of when these models need to be transformed and replaced by other more appropriate models (Allen (94) describes this as 'the "learning" process of trying to "model" the situation').

Morin (1992) also supports this provisional view of knowledge. He argues that a recognition of complexity gives us licence to challenge assertive positions, as is

[15] See Sects. 2.6.1 and 2.6.2.

clear from his statement that '[i]t is because official certainty has become uncertain that official intimidation can let itself be intimidated in turn' (11). In a later text, Morin (2007: 21) discusses the implications that this provisional stance holds for modelling our complex realities. According to him:

> There is no science of science, and even the science of science would be insufficient if it did not include epistemological problems. Science is a tumultuous building site, science is a process that could not be programmed in advance, because one can never program what one will find, since the characteristic of a discovery is in its unexpectedness. This uncontrolled process has led today to the development of potentialities of destruction and manipulation, which must bring the introduction into science of a double conscience: a conscience of itself and an ethical conscience.

In developing an ethical conscience, the scientist must learn to take responsibility for her decisions. Our models are the outcome of framing strategies, which, as Allen (2000) notes, represent possible choices amongst others. Each choice gives rise to 'a different spectrum of possible consequences, different successes and failures, and different strengths and weaknesses' (102)—most of which cannot be known beforehand. Therefore, although we must exercise vigilance when choosing our strategies, we should also recognise that no matter how carefully we consider and reflect upon these actions, they may turn out to be a mistake (Preiser and Cilliers 2010). Preiser and Cilliers (274) summarise the implication of this point as follows:

> acknowledging that values and choice are involved does not provide any guarantee that good will come of what we do. Complexity tells us that ethics will be involved, but does not tell us what that ethics actually entails. The ethics of complexity is thus radically or perpetually ethical… We do not escape the realm of choice.

In summary, the epistemology of complexity cannot be construed as a value-free programme, aimed at establishing *a priori* laws and rules through reductive reasoning. This is because such a conception represents not only the negation of complexity, but also 'destroys the autonomy of one who seeks' (Morin 2007: 28). Engaging with the problematic of complexity therefore necessitates what Morin (28) calls 'an autonomous strategy' which obliges us 'in the field of action… to reconsider our decisions like bets and incites us to develop an adequate strategy to more or less control the action.'

Although daunting, Cilliers (2001) argues that the inability to represent knowledge objectively, or the inability to close the gap between a model and that which is being modelled, should 'serve as a creative impulse that continually challenges us to transform our models, not as a reason to give up' (138). Morin (1992: 10) also notes the potential virtues of 'ignorance, uncertainty, and confusion', in arguing that these virtues guard against a 'mutilating simplification' (10). They thus keep the critical and reflexive impulse alive. In his words:

> Uncertainty becomes viaticum: doubt of doubt gives rise to a new dimension, the dimension of reflexivity; the doubt by which the subject questions the conditions of emergence of its own thought constitutes henceforth a potentially relativist, relationist and self-knowing thought (10).

4.5 The Question of the Foreigner

The preceding analysis provided insight into the manner in which meaning is formed and transformed. The reader's attention was also drawn to the inevitable ethical dimension that characterises the process of meaning formation in a complex and immanent world. The ethics of living denotes, amongst other things, a critical awareness of how our knowledge practices come about, the status of these practices, and the consequences that they potentially hold for us and for others. Up until this point in the analysis, few examples have been used in order to substantiate the theoretical discussion. However, at this juncture I wish to turn to the question of the foreigner in order to further interrogate the ethical implications that the view of meaning or knowledge espoused in this chapter holds. Although there are other examples that could also serve my purpose, the question of the foreigner is particularly apt because it poignantly illustrates the ethical challenges inherent to modelling and classifying. As mentioned in the introduction to this work, this example also represents one of the foremost political challenges of our time, and it has received significant attention in the work of Derrida and Nancy. A last reason for choosing the example of the foreigner is because a couple of years ago, I had an experience that revealed my own prejudice to foreigners, and that highlighted the deep ethical issues that the question of, and responses to, the foreigner raise. I begin the analysis with a short anecdote of this experience, which I then analyse in terms of Derrida's exploration of both foreignness and hospitality. The significance of this analysis for the ethics of knowledge will also be highlighted.

4.5.1 A Foreigner's Call

In May 2012, my family and I went away for a weekend to a game farm in the Waterberg to celebrate my mother's birthday. Although the Waterberg is only a two hours drive from Pretoria, the sparsely populated bushveld and the vast open skies mean that one feels as if one has travelled to a world far away from civilization. We spent a glorious weekend lazing about, going on game drives, and relaxing around a fire under the stars.

At about 10:00 p.m. on the last evening, we were all lying in bed reading when suddenly we became aware of a man crying out into the night. His words were indistinct and his voice sounded tortured. My first thought was that the man was drunk. Although I felt some apprehension, I was comforted by the fact that the sound came from far away. Over the course of the next while however the man's screams became progressively louder and closer, and it was clear that he was making his way to our cottage.

At this point, I, more so than any of my family members, was overcome by a deep sense of fear—especially for the life of my then baby daughter, who was asleep in our room. Farm murders are not uncommon in South Africa, and my family and

I were completely defenceless: we were stuck in the middle of the bush, in a cottage with flimsy doors and windows, and with nothing to protect us. I took action and began dialling the emergency numbers in the visitor's information book. After frantically dialling several numbers, I finally got hold of the owner of the game farm, and explained the situation to him, begging him to come to assist us. His house was about half an hour from where we were staying, and it was clear that the howling man—who by this stage sounded as if he was just down the road—would reach us before he would. He told us to hang tight. He would contact the police, assemble some farm workers, and be at our cottage as soon as possible. There was nothing to do but to await our fate.

A couple of minutes later, the howling man was knocking at the windowpane, shouting incoherently in a language that we did not understand. He did not however try to break down the door or knock out a window, as I had anticipated. Gradually, we became aware of one word that we could understand: help. My fear was beginning to subside, but I forbade anyone to leave the cottage. We indicated to the man that he should wait by the fire outside and that help was on the way. He did as he was told and stopped shouting.

Eventually, the owner of the farm arrived with some of his workers (we had again contacted him to tell him that it did not look as if we were in mortal danger, and that it seemed unnecessary for the police to come as well). As soon as the farm men arrived, my father and brother ran out to find out what was going on. They later reported that seated at the fire was the man and his wife. The wife had all but stopped breathing and was evidently in dire need of medical attention.

The next day the farm owner phoned us in order to inform us about the events that had transpired the previous evening: as they were driving away from our cottage, the man started screaming incoherently again. It had apparently taken a long time to figure out what he was trying to say, but it finally emerged that he was the foreman of the farm on the neighbouring property. His wife had fallen ill the previous day, and had received some *muti* from the local medicine man. She had combined the *muti* with alcohol and had lost consciousness. Her husband realised that she urgently needed medical help, but he had no phone and could not contact anyone. At this stage, he started calling out for someone to help him and had started the long walk towards our cottage. He and his wife also had a small baby, who he could not leave alone at home, so he was forced to carry both his unconscious wife and his baby. About halfway through his journey, the weight of both his wife and baby became too much for him, and he realised that he would have to leave one of them behind. He decided to hide his baby in the bushes—despite the fact that the game farm was home to a number of animals of prey—and continued on his journey with only his wife.

According to the farm owner, it had taken them some time to find the baby in the dark and cold night, but eventually they came across the little human bundle, who—mercifully—seemed unharmed. Mother and child were driven to the hospital, and both received medical attention and were able to return home a day later.

The experience impacted heavily on my family and me, and the mood on the return journey to my parents' house in Pretoria was sombre and reflective. I was deeply ashamed of myself for the way that I had reacted the previous night, especially because it had never once occurred to me that the howling man was in need of a fellow human being's help. I had assumed—and never doubted despite evidence to the contrary—that he was a mad person intent on doing harm to my family and me. In other words, I had treated this poor and suffering man as a hostile stranger, and never once swayed in my opinion of him.

4.5.2 *Derrida: The Foreigner and the* **Aporia** *of Hospitality*

My experience raises a number of themes related to foreignness (i.e. that which/who lies beyond our knowledge) that Derrida also touches on in his analysis of the issue. Important in this regard is the manner in which the foreigner disrupts the logos and the response of classifying the foreigner as mad, the definition of the foreigner and how it is related to questions of inclusion and exclusion, the issue of language and how it translates into violence, and the response to the foreigner and the *aporia* of hospitality.

Before investigating these issues, it should be noted that Derrida's treatment of the foreigner and hospitality (as with his treatment of a range of key concepts such as law, justice, friendship, the gift, testimony, cosmopolitanism, and forgiveness) can be described as a type of conceptual genealogy. This is very similar to Foucault's effective history in that Derrida 'selects a concept from what he always describes as 'the heritage'—let's call it the dominant Western tradition—and then proceeds, via an analysis that is at once historical, contextual, and thematic, to bring out the logic of that concept' (Critchley and Kearney 2001: viii–ix). Derrida traces the related questions of the foreigner and hospitality 'back to the Greco-Roman world, through the Judeo-Christian tradition, and to the political philosophies of Kant and Hegel' (Westmoreland 2008: 1). I shall only touch on Derrida's treatment of the subject at the hand of Plato's dialogues, and Kant's (2006) text, titled *Toward Perpetual Peace*.

In his reading of Plato's *Sophist,* Derrida (2000a) emphasises the threat that the foreigner or the *Xenos* holds for the paternal authority of the city. The foreigner—the sophist—comes from outside the city and does not subscribe to its authority. Indeed, the foreigner is a foreign son, whose father is unknown. He is severed from the authority and the house of his father, and carries the unbearable question of parricide as a possible counter to the paternal logos of the Greek father Parmenides. The foreigner threatens to destabilise the conceptual hierarchy according to which the city is organised because he does not share in the culture, traditions, and ordering of the city. He must therefore be disarmed. The foreigner is aware of the threat that he holds for the city, and as such he is fearful of being declared 'mad, *manikos*, a nutter, a maniac "who is upside down all over… a crazy person who reverses everything from head to toe, from top to bottom, who puts all his feet on his head, inside out, who walks on his head"' (11).

As noted by Plato in the *Statesman*, apart from declaring him mad, there is a second way in which to disarm the foreigner, and that is through the political apparatus of the state. Politics enables the instantiation of a pact that links the foreigner to the city, its knowledge, and its laws. This happens to the extent that foreigners become 'subjects in law, to be questioned and liable, to have crimes imputed on them, to be responsible, to be equipped with nameable identities and proper names' (23). The identity accorded to the foreigner also serves to distinguish the foreigner from the citizen. In so doing, the foreigner's identity grounds the identity of the citizen ('I am a citizen and therefore different to you, a mere foreigner'). For this reason, Derrida writes that '[a] proper name is never purely individual' (23). The classification of the foreigner therefore places him 'inside the law, under the law, essential to the law' (Westmoreland 2008: 2).

By extending a contract to the foreigner, rights are conferred to the foreigner, which also constitutes a show of hospitality to the foreigner. Indeed, during his plea, Socrates complained that the court did not even treat him as a foreigner. Had this been the case, the court would be more tolerant of the fact that he was unfamiliar with the legal jargon (Derrida 2000a). Yet, as Derrida reminds us, the classification of the foreigner under the law (conferring upon him a proper name) both enables and limits (even prohibits) the hospitable relation with the foreigner. This is because hospitality is caught up in an economy of exchange, whereby the foreigner is (to varying degrees) assimilated by the logos. According to Samuel Weber (1992: 251), 'the problem of the law can be formulated… as that of translating an unconditional promise [of hospitality] into a conditional one.'

The problem of language aptly illustrates the paradox of hospitality. The foreigner seeking asylum is expected to renounce his foreignness in order to ask for hospitality under a system and in a language that is not his own. According to Derrida (2000a: 16), the fact that he is expected to master this foreign system and language in order to receive the right of hospitality constitutes 'the first act of violence'. This further begs the question as to whether he can still be called a foreigner and be shown hospitality once he is violently assimilated into the system. Derrida (16–17) describes this problematic as follows:

> That is where the question of hospitality begins: must we ask the foreigner to understand us, to speak our language, in all the sense of this term… before being able and so as to be able to welcome him into our country? If he was already speaking our language… if we already shared everything that is shared with language, would the foreigner still be a foreigner and could we speak of asylum or hospitality with regard to him? This is the paradox that we are going to see become clearer.

Derrida also examines the conditional nature of hospitality at the hand of Kant's text *Perpetual Peace*. Derrida (2000b: 3) writes that in this text, Kant is interested in:

> defining the conditions of a cosmopolitan right, of a right the terms of which would be established by a treaty between states, by a kind of UN charter before the fact, and one of these conditions would be what Kant calls universal hospitality, *die allgemeine Hospitalität*.

Important in this regard is that the conditions put forward are without geographical limit, even though Kant recognises that all regions are not unconditionally accessible to everyone (due, primarily, to the institution of states) (Derrida 2001). Given the provisos of unconditional geographical reach, but conditional political access, Kant puts forward two conditions for universal hospitality. The first condition is that hospitality is limited to the right of visitation:

> in this context, *hospitality*... means the right of a stranger not to be treated in a hostile manner by another upon his arrival on the other's territory... It is not the *right of a guest* that the stranger has a claim to... but rather a right to visit, to which all human beings have a claim, to present oneself to society by virtue of the right of common possession of the surface of the earth (Kant 2006: 82).

The second condition is that the rights of residence and hospitality (and, conversely, infringements of hospitality) are subject to law, and contractual agreements between states. In drawing on examples of the inhospitable behaviour of certain groups of coastal and desert people towards seafarers and nomadic groups respectively, Kant (82) contextualises and explicates this condition as follows:

> The right of hospitality, that is, the right of foreign arrivals, pertains, however, only to conditions of the possibility of *attempting* interaction with the old inhabitants.—In this way, remote parts of the world can establish relations peacefully with one another, relations which ultimately become regulated by public laws and thus finally bring the human species ever closer to a cosmopolitan constitution.

Derrida praises Kant's rigour in formulating the law of cosmopolitanism and he also acknowledges that the conditions of perpetual peace must necessarily be institutionalised in order to serve as universal guidelines for the future. However, Derrida also warns that Kant's argument is constructed on the assumption that we *know* what hospitality is (Derrida 2000b).

In addition to a conditioned, and hence a known form of, hospitality, Derrida identifies an unconditional hospitality, which 'is... an intentional experience which proceeds beyond knowledge toward the other as absolute stranger, as unknown, where I know that I know nothing of him' (8). Derrida thus draws a distinction between the foreigner and the stranger (the other[16]), who cannot be assimilated, and who therefore does not have a name or patronym (Derrida 2000a). Derrida further argues that this view of hospitability as absolute implies a break with hospitality understood in the ordinary sense of a right or pact, for the very reason that it 'is structured as a universal singularity, without imperative, order, or duty' (Westmoreland 2008: 3).

Unconditional hospitality requires that we welcome the stranger who arrives without invitation, and who is defined as an absolute, inassimilable other. Unconditional hospitality therefore 'relies upon the deafening silence between the *ipse* and the other' (Derrida 2000a: 5). Unconditional hospitality also constitutes a willingness to transgress established boundaries, and to be taken hostage in your own home. This is because the arrival of the stranger dethrones the host by denying her access to the rules and laws by which she exercises hospitality (in that these rules and laws are already directed towards a certain conception of the foreigner).

[16] This reference to the other recalls Levinas's philosophy, which will be explored in Sect. 5.2.

In challenging the laws and logic according to which we structure our lives, the stranger thus also challenges the very identity of the host and of being. The stranger, in other words, puts me in question (Derrida 2000a). Derrida (2002b: 364) writes that '[h]ospitality is the deconstruction of the at-home; deconstruction is hospitality to the other'. Yet this experience also cannot last. Absolute hospitality can only take place in an instant, before being dialecticised and taken up into knowledge (Derrida 2000b), and before imposing the limits and conditions necessary to secure the *ipse* as master of the house (Westmoreland 2008).

Similar to the nature of justice and the gift,[17] true hospitality is thus non-formalisable—in Derrida's (2000b: 8) words, 'hospitality gives itself, and gives itself to thought beyond knowledge'. For this reason, Derrida argues that hospitality (1999b: 50) serves as the foundation or 'the whole and the principle of ethics'. This is true in so far as ethics is understood as an experience of the impossible, in that it is without concept and without law (Bernasconi 1987; Derrida 1978). In this interpretation, ethics cannot simply be reduced to a question of epistemology; rather, it constitutes as engagement with radical alterity. Even though absolute hospitality renders the law of hospitality as always already corrupted, and even though knowledge of absolute hospitality is unattainable, Peggy Kamuf (2006) argues that this is not a cause for despair. Rather, the *aporia* of hospitality requires that we think together both horns, for the reason that this double-thinking is precisely what saves our practices. She explains as follows:

> To think the unconditionality of [limit] concepts is not at all to remove thought from the practical experiences we wish to call hospitality, gift, forgiveness, or justice. On the contrary, this thinking registers the very desire to go on calling to these names for that which remains impossible as present experience (207).

4.5.3 *Hospitality-Hostility*

My reaction to the stranger encountered in the bushveld betrayed my failure to think the unconditionality of hospitality. In that specific instant, the political apparatus of the state that allows for classification and ordering failed, for the reason that we were in the middle of nowhere and thus could not call upon this apparatus. In other words, in that instant, the man could not be definitively identified as an unemployed shack-dweller, a Zimbabwean immigrant, a Sotho farm worker, or any of the other categories that—under other circumstances—could be drawn upon in order to confer some type of identity upon him. We thus perceived the man as a nameless stranger—who, through his persistent and ever-louder calling, was intent on transgressing the definitive boundary between the inside of our family cottage and the dark outside; and who, through virtue of his strange behaviour, both unsettled and threatened the peaceful order of our household. This stranger confronted us in a language (both verbal and non-verbal) that we did not understand. Aside from

[17] See Sect. 3.4.2.

uttering the word 'help', he was unable to translate his request for help in terms that were familiar enough to make me feel comfortable to open the door for him. In this instance, the *Xenos* became the madman, who remained threatening and untameable (even when the moment of threat had passed).

Although this anecdote refers to an isolated experience that occurred between a family and one stranger, it does reveal the extraordinary challenge that absolute hospitality sets before us: hospitality requires of us to be prepared to welcome a stranger, without invitation, and without qualification. It asks of us to be prepared to allow the other to violate our home and our identity to the extent that the stranger imposes upon our hospitality, and thus transforms our identity of host or master of the house to that of the stranger's hostage.[18] The stranger, in other words, challenges the sovereignty that I have over myself and my home, and which I exercise by 'filtering, choosing, and thus by excluding and doing violence' (Derrida 2000a: 55). Derrida (53–54) argues that the consequence of this is that:

> one can become virtually xenophobic in order to protect or claim to protect one's own hospitality, the own home that makes possible one's hospitality… Anyone who encroaches on my "at home", on my ipseity, on my power of hospitality, on my sovereignty as host, I start to regard as an undesirable foreigner, and virtually as an enemy.[19]

Derrida further argues that this type of xenophobia is exacerbated in an age where technological advances are continuously shrinking the private sphere. To this one can add that globalisation at large, coupled with the displacement of large groups of people due to persistent conflict in various parts of the world, also threaten the ability of states to protect their own hospitality. This fuels the fires of xenophobia. In order to counteract these xenophobic attitudes, and the concomitant perversion of the law of hospitality, it is necessary to continuously and consciously reflect on the demands of absolute hospitality. Yet, Derrida (2001: 22–23) also states that this reflection must be geared towards improving and transforming the laws of conditional hospitality, otherwise '[t]he unconditional law of hospitality would be in danger of remaining a pious and irresponsible desire, without form and without potency, and of even being perverted at any moment'.

4.6 The Ethics of Modelling

The paradox of the foreigner/stranger is another variation of the problem encountered in the previous chapter, namely how to speak of a general economy from the perspective of a restricted economy or an economy of meaning. In order for a stranger to be a stranger, her strangeness cannot be assimilated into the system.

---

[18] This point will be returned to in Sect. 5.5.

[19] In this regard, it is interesting to note that the word *hospitality* has Latin roots; and that, in Latin, *hospitality* 'allows itself to be parasitized by its opposite, "hostility," the undesirable guest [*hôte*] which it harbours as the self-contradiction in its own body' (Derrida 2000b: 3). This description accords with Nancy's view of the intruder, which is discussed in Sect. 5.5.

However in order to recognise the foreigner as someone different to me, I need to assimilate or homogenise the stranger, at least to the extent that she becomes understandable to me. This paradox can also be explained at the hand of Human and Cilliers' (2013) discussion on the difference between heterogeneity and difference. According to them, difference can only be recognised in terms of a common frame that allows us to make discriminations on the basis of this given perspective. Heterogeneity, to recall, is that which is non-sensical from the perspective of a given model. Following Bataille (1989), Human and Cilliers argue that we can only understand heterogeneity by including it as a difference within a model. However, to do so destroys the very heterogeneity that we seek to understand, whilst simultaneously producing new forms of heterogeneity that lie outside this newly-accepted set of differences. In terms of our example, this means that to accept a given stranger as a foreigner leads to a re-articulation of the boundaries of the system, thereby making it possible for new forms of strangeness to emerge.

The example of the foreigner is a very concrete illustration of the challenges inherent to knowledge generation in general, specifically as concerns the assimilation of any variable into the logic of a given conceptual system. If a given variable cannot be translated or reduced to the logic of the system, then it is relegated to the outside of the system, where it is either forgotten or treated in hostile terms (recall that the stranger is suspected of parricide). The specific challenge of modelling is thus that of recognising and working productively with both the necessity of modelling, and the inherent limitations to modelling. In this final section, a summary of insights gleaned in this chapter is provided. These insights pertain to the nature of modelling, and to the possibility and the nature of intervention.

In the introduction, it was stated that modelling is necessary in order to gain knowledge of complex phenomena. However, the act of modelling not only reduces the complexity of the system under study, but also distorts our understanding of the system, and therefore introduces more complexity (Luhmann (1995: 471) refers to this as 'hypercomplexity'). The reasons for this are premised on the normative dimension of modelling. Firstly, modelling constitutes a choice amongst alternatives. This is because, as Derrida explains, the text is complicated (in our language, context is complex). This means that there is too much determinacy, and hence too many (and often contradictory) variables to be included in a single model. Meaning is therefore always politicised in that every act of modelling creates hierarchies and exclusions, whereby some variables are privileged, others are marginalised, and others are completely excluded from the model.

Derrida's treatment of Plato's analysis of the foreigner aptly illustrates the implications that modelling holds for issues of inclusion and exclusion and assimilation and non-assimilation. In this example, the logos belongs to the citizen (who is male); women, children, and foreigners represent the marginalised terms; and the stranger is excluded from the city. However (and as mentioned above), even though the foreigner is to a certain extent recognised within the schema of the city, it is important to realise that recognition comes at the price of denying the foreigner's foreignness (his language, his culture, etc.). What this example implies is that the status that we accord to the various variables within a system is influenced by the meta-assumptions upon which the model is based.

Secondly (and following from the above), it is important to recognise that we never approach the act of modelling from a clean slate. Our models are premised on our sensory apparatus and our physical and cognitive resources (including our individual judgements, preferences, biases, opinions, etc.). As further shown by both Foucault and Derrida, our understanding is always already informed by certain epistemic (Foucault) or philosophical (Derrida) frames, i.e. by a type of meta or theoretical context that lends a certain interpretation to these specific contexts. Moreover, and as illustrated by Foucault, in many cases we are not even aware of these meta-assumptions. This, Derrida argues, leads to the naturalisation of our models. In terms of the example of foreignness, this may very well serve as a basis for the rationalisation of xenophobia, to the extent that it is deemed natural to view the foreigner as inferior and undeserving of consideration.

Thirdly, both this meta-context and our specific contexts are dynamic. This is because present-day rules of formation or philosophical predicates, as well as the meaning of specific concepts, are informed by a rich, complex, and non-linear history. This is again illustrated in both the work of Foucault and Derrida, to the extent that they draw on genealogical investigations of what Foucault calls *effective history*, and what Derrida refers to as *the heritage* in order to better understand the meaning and value of present-day concepts, including (in Derrida's case) the concept of foreignness. Meaning is thus also contingent on a temporal dimension, as traces of past meanings inform our understanding of present-day concepts. Therefore, apart from being informed by a given model (which, in itself, constitutes a partial and distorted understanding of reality), the meaning that we accord to a variable or concept is also the product of a complex and impure history.

All three of the above points underscore Derrida's view that 'meaning is context-bound, but context is boundless' (Cullers 1983: 123). There is thus an unbridgeable abyss that exists between meaning and context, and this abyss is both the product of complexity, and also generates complexity. The manner in which meaning is constituted necessitates an understanding of complexity as a critical enterprise, and also inaugurates the possibility of intervention (as a necessary condition of the critical enterprise). In other words, intervention is necessary precisely because meaning is incomplete and impure (in other words, no concept can fully correspond with itself), and yet we employ concepts as if they exhaust meaning.

Apart from fact that the logic of meaning-formation necessitates intervention, there are also substantive reasons for intervention. In the social realm specifically, intervention is needed because the way in which meaning is formed materially affects the lives of people. In my bushveld example, my interpretation of the stranger as a madman, instead of as a fellow human being in need of help, could have potentially cost the life of his wife or child (had there not been other agents to intervene in the situation). Intervention therefore not only follows from the formal structure of the argument, but is also informed by a social-political impetus.

The fact that intervention is necessary does not yet shed light on the nature of responsible intervention. Whereas Foucault views intervention as a form of individual resistance against the workings of bio-power, Derrida develops an *aporetic* understanding of deconstruction-as-intervention. This is fuelled by the desire to remain sensitive to the political and ethical consequences that meaning carries. The *aporia* that Derrida works with can variously be expressed as the tension between assimilation and non-assimilation, action and reflection, practice and theory, or institutions and philosophy. In each case, the two horns of the *aporia* are both separated and tied together by a complex logic, which requires that we think these horns together.

The concrete example discussed in this chapter in order to illustrate the workings of the *aporia* was the example of the relation between the law of hospitality and unconditional hospitality, which represent the economic and aneconomic dimensions of hospitality respectively. To act responsibly and, moreover, to model responsibly, means that we must remember that the foreigner remains a stranger, even when we are in the process of reducing his strangeness. Importantly then—and as previously argued—the epistemology of complexity concerns not only the nature of models, but also the process of modelling and the attitude of the modeller. The latter two features specifically introduce an awareness of the ethics of modelling. The ethics of modelling serves as impetus to treat our models as provisional and to remain vigilant to the fact that these models are in constant need of intervention and transformation. The ethics of modelling should thus be understood in terms of a recursive modality.

Apart from the consequences that the inexhaustibility of meaning holds for the ethics of modelling, unconditional hospitality and questions concerning identity also draw attention to Derrida's ethics of radical alterity. As with Levinas's view of ethics, which will be addressed in the following chapter, Derrida appeals to otherness as the condition for the possibility of an ethics that cannot be appropriated in (or as) meaning, but that is nevertheless necessary for meaning formation and transformation. In the following two chapters, this ethics of radical alterity (which also characterises a complex view of ontology) will be addressed in more detail.

4.7 Conclusion

This chapter focused on the nature of meaning as understood from philosophical positions that are sensitive to complexity, specifically Foucault's and Derrida's positions. A number of insights pertaining to the nature of modelling and intervention came to the fore, and these insights were explored in terms of the ethics that they introduce to the acts of modelling and intervention. Ultimately, to model and to intervene responsibly requires a complex thinking because the modeller is forced to contend with the *aporia* that defines meaning and that holds the consequence that models cannot accurately, or completely, describe the systems and contexts that they represent. The paradox of the foreigner, which gives rise to the *aporia* of hospitality, adequately illustrates why we cannot simplify our thinking or forego the imperative of remaining engaged with the *aporetic* logic that defines meaning formation.

Despite the focus on meaning generation, a number of insights pertaining to the nature of identity also emerged during the course of the analysis. In the introduction it was stated that the subject is constituted within a network of knowledge, which means that the subject-object divide is untenable. This insight was supported by the analysis presented of Foucault's position, in which he demonstrates that the human is simultaneously a subject (hence the source of knowledge) but also an object (hence the subject of knowledge). The body of knowledge that we produce therefore produces us, and this reciprocal process gives rise to the workings of bio-power. The introduction to Derrida's deconstructive philosophy also holds consequences for our view of identity in so far as identity is shown to be constituted within conceptual hierarchies, wherein certain identities are privileged over others. The identity of the foreigner, for example, is determined within the political apparatus of the state, and is defined relative to the identity of the citizens. Through means of deconstruction, Derrida is however able to show that, instead of being oppositionally-determined, identities defer to one another in a constant exchange of meaning. This is supported by Derrida's analysis of the interchange of properties between host and hostage (which will be explored in more detail in the next chapter).

What these insights regarding identity construction illustrate, is that questions concerning meaning and knowledge are inseparable from identity-questions. This adds yet another layer of complexity to the analysis, and even though these dimensions are treated separately in this work for the sake of expediency and clarity, it is important to bear in mind that such a separation represents an artificial demarcation. This said, focusing on identity and alterity raises further issues that have, as yet, not been considered in this analysis, and it is to these issues that we turn in the following chapter.

References

Allen, P. 2000. Knowledge, ignorance, and learning. *Emergence* 2(4): 78–103.

Bataille, G. 1989. The notion of expenditure. In *Visions of excess: Selected writings, 1929–1939,* ed. A. Stoekl, trans. A. Stoekl, C.R. Lovitt, and D.M. Leslie Jr., 116–129. Minneapolis: University of Minnesota Press.

Bernasconi, R. 1987. Deconstruction and the possibility of ethics. In *Deconstruction and philosophy: The texts of Jacques Derrida*, ed. J. Sallis, 122–139. Chicago: University of Chicago Press.

Bélanger, J.L.R. 1992. Translator's introduction. In *Method: Towards a study of humankind vol. 1: The nature of nature,* E. Morin, trans. and intro. J.L.R. Bélanger, xiii–xiv. New York: Peter Lang Publishing.

Cilliers, P. 2000. Knowledge, complexity, and understanding. *Emergence* 2(4): 7–14.

———. 2001. Boundaries, hierarchies and networks in complex systems. *International Journal of Innovation Management* 5(2): 135–147.

———. 2005. Complexity, deconstruction and relativism. *Theory, Culture & Society* 22(5): 255–267.

———. 2010. A provisional ethics of complexity, unpublished.

———. 2016. *Critical complexity: Collected essays,* ed. R. Preiser. Berlin: De Gruyter/Ontos.

Critchley, S., and R. Kearney. 2001. Preface. In *On cosmopolitanism and forgiveness*, ed. J. Derrida, vii–xii. New York: Routledge.

Culler, J. 1983. *On deconstruction: Theory and criticism after structuralism*. London: Routledge.

Derrida, J. 1976a. *Of grammatology*. Trans. G. Spivak. Baltimore: Johns Hopkins University Press.

———. 1976b. Entre crochet (an interview). *Digraphe* 8: 97–114

———. 1978. Violence and metaphysics. In *Writing and difference*, 79–153. Trans. A. Bass. Chicago: University of Chicago Press.

———. 1979. The parergon. Trans. C. Owen. *October* 9: 3–41.

———. 1980. The law of genre. Trans. A. Ronell. *Glyph* 7: 202–229.

———. 1982. The conflict of the faculties. Trans. R. Rand and A. Wygant. In *Knowledge of language and of inquiry*, ed. M. Riffaterre. New York: Columbia University Press.

———. 1988a. Letter to a Japanese friend. In *Derrida and différance*, ed. D. Wood, 1–5. Evanston: Northwestern University Press.

———. 1988b. Afterword. In *Limited Inc*, ed. G. Graff, trans. S. Weber, 111–160. Evanston: Northern Western University Press.

———. 1999a. Hospitality, justice and responsibility: A dialogue with Jacques Derrida. In *Questioning ethics: Contemporary debates in philosophy*, ed. R. Kearney and M. Dooley, 65–68. London/New York: Routledge.

———. 1999b. *Adieu*. Trans. P.-A. Brault and M. Naas. Stanford: Stanford University Press.

———. 2000a. Foreigner question: Coming from abroad/from the foreigner, fourth seminar (January 10, 1996). In *Of hospitality*, on invitation from A. Dufourmantelle. Trans. R. Bowlby, 3–74. Stanford, CA: Stanford University Press.

———. 2000b. Hostipitality. Trans. B. Stocker with F. Morlock. *Angelaki: Journal of the Theoretical Humanities* 5(3): 3–18.

———. 2001. On cosmopolitanism. In *On cosmopolitanism and forgiveness*, 3–24. Trans. M. Dooley. New York: Routledge.

———. 2002a. Negotiations. In *Negotiations: Interventions and interviews, 1971–2001*, ed. and trans. E. Rottenberg, 11–40. Stanford: Stanford University Press.

———. 2002b. Hospitality. In *Acts of religion*, ed. and trans. G. Anidjar, 356–420. New York: Routledge.

Dreyfus, H.L., and P. Rabinow. 1983. *Michel Foucault: Beyond structuralism and hermeneutics*, 2nd ed. Chicago: University of Chicago Press.

Foucault, M. 1967. Nietzsche, Freud, Marx. In *Nietzsche*. Trans. J. Anderson and G. Hentzl. Paris: Cahiers de.

———. 1973. *Madness and civilization: A history of insanity in the age of reason*. Trans. R. Howard. New York: Vintage/Random House.

———. 1975. *The birth of the clinic: An archaeology of medical perception*. Trans. A.M. Sheridan. New York: Vintage/Random House.

———. 1979. *Discipline and punish: The birth of the prison*. Trans. A. Sheridan. New York: Vintage/Random House.

———. 1990. *The history of sexuality*. Vol. 1, *An introduction*. Trans. R. Hurley. New York: Vintage/Random House.

———. 1984. Nietzsche, genealogy, history. Trans. D.F. Bouchard and S. Simon. In *The Foucault reader: An introduction to Foucault's thought*, ed. P. Rabinow, 76–100. London: Penguin.

———. 2002a. *The order of things: An archaeology of the human sciences*. Trans. unspecified. New York: Routledge Classics.

———. 2002b. *The archaeology of knowledge*. Trans. A.M. Sheridan Smith. London: Routledge Classics.

Heidegger, M. 1962. *Being and time*. Trans. J. Macquarrie and E. Robinson. New York: Harper and Row.

Human, O., and P. Cilliers. 2013. Towards an economy of complexity: Derrida, Morin and Bataille. *Theory Culture Society* 30(5): 24–44.

Hurst, A. 2004. Derrida's quasi-transcendental thinking. *South African Journal of Philosophy* 23(3): 244–266.

Kamuf, P. 2006. From now on. *Epoche* 10(2): 207.

Kant, I. 2006. *Toward perpetual peace and other writings on politics, peace, and history*, ed. P. Kleingeld, trans. D.L. Colclasure. New Haven: Yale University Press.

Luhmann, N. 2000. Why does society describe itself as postmodern? In *Observing complexity: Systems theory and postmodernity*, ed. W. Rasch and C. Wolfe, 35–49. Minneapolis/London: University of Minnesota Press.

Luhmann, N. 1995. *Social systems*. Trans. J. Bednarz, Jr. (with D. Baecker). Stanford: Stanford University Press.

Markos, L. 2012. Lecture on structuralism [online]. http://www.youtube.com/watch?v=bcR-4y3Npqc. Accessed 1 August 2014.

Morin, E. 1992. *Method: Towards a study of humankind*, vol. 1, *The Nature of Nature*, trans. and intro. J.L.R. Bélanger. New York: Peter Lang.

———. 2007. Restricted complexity, general complexity. Trans. C. Gershenson. In *Worldviews, science and us: Philosophy and complexity*, ed. C. Gershenson, D. Aerts, and B. Edmonds, 5–29. Singapore: World Scientific.

———. 2008. *On complexity*. Trans. S.M. Kelly. Cresskill: Hampton Press.

———. 1977. *La Méthode, v. 2: La Nature de la nature*. Paris: Seuil.

Oreskes, N., K. Shrader-Frechette, and K. Belitz. 1994. Verification, validation, and confirmation of numerical models in the earth sciences. *Science* 263(5): 641–646.

Nietzsche, F. 1989. *On the genealogy of morals*. Trans. W. Kaufmann and R.J. Hollingdale. New York: Vintage Books.

Preiser, R., and P. Cilliers. 2010. Unpacking the ethics of complexity: Concluding reflections. In *Complexity, difference and identity*, ed. P. Cilliers and R. Preiser, 265–287. Dordrecht: Springer.

Rosen, R. 2005. *Life itself: A comprehensive inquiry into the nature, origin, and fabrication of life*. Columbia: Columbia University Press.

Rorty, R. 1978. Philosophy as a kind of writing: An essay on Derrida. *New Literary History* 10: 141–160.

Weber, S. 1992. In the name of the law. In *Deconstruction and the possibility of justice*, ed. D. Cornell, M. Rosenfield, and D.G. Carlson, 232–257. New York: Routledge.

Westmoreland, M.W. 2008. Interruptions: Derrida and hospitality. *Kritike* 2(1): 1–10.

Chapter 5
Complex Identity and Ethical-Political Responsibilities

> *How then can the pour-soi (for-self) be transformed into the pour-tous (for-all), while remaining frenetically pour-soi? We can begin to understand that from the moment one living being becomes an existential exigency for another; this exigency immediately creates, in fact, a solidarity and a complementarity of the one in relation to the other.*
>
> *-Edgar Morin (1980), La Méthode, v. 2: La Vie de la vie*

Abstract

Philosophical complexity and post-structuralism both offer a challenge to the Cartesian humanist subject and the predicates of subjecthood that this view presupposes (including a strong view of agency, intentionality, rationality, and causality). In substantive terms, this challenge results in a so-called liquidated subject, wherein the notion of the subject no longer corresponds with any fixed or signified content but is instead characterised as a decentred and complex construction.

In this chapter, the deconstruction of the humanist subject and the traditional predicates of subjecthood is undertaken at the hand of Levinas's understanding of the Other; Derrida's work on the subject, animals, and eating; and, Nancy's reading of Heidegger's Dasein from the perspective of Mitsein. The insights that these analyses yield are critically compared to Cilliers' complex view of identity, in which the self becomes over time in a network of relations with others. In so doing, a tentative portrait of the liquidated subject emerges, and attention is drawn to the urgent need to revise our traditional understanding of ethics and responsibility.

The consequences that this complex view of identity holds for understanding and relating to the self and to the other are also explored at the hand of the example of encountering the stranger (Levinas, Derrida) or the intruder (Nancy).

M. Woermann, *Bridging Complexity and Post-Structuralism*,
DOI 10.1007/978-3-319-39047-5_5

5.1 Introduction

As stated in the conclusion to the previous chapter, we cannot separate what we know from who we are; and, just as we cannot have an objective idea or an accurate meta-model of the world, so too we cannot have a fixed *a priori* view of the self. The idea of a fixed self has its origins in Cartesianism. The Cartesian subject, which is introduced in Descartes' (1960) *Discourse on Method*, marks the birth of the modern, philosophical history of the subject. Of specific importance is 'The Second Meditation,' wherein Descartes contemplates the nature of the human mind, and the fact that it is easier to know the mind than the body.

In 'The second meditation', Descartes conducts a thought experiment in order to help him gain both certainty of, and clarity on, the nature of his existence. The self-evidence of the subject is however not presupposed in his analysis. In fact, Descartes begins 'The first meditation' by assuming that everything he sees is false. His starting point is thus that nothing that his deceptive memory has represented to him has ever existed. He thinks himself without his senses, body, figure, place, extension, and movement since he contends that these 'are all fictions of [the] mind' (107). The Cartesian subject is therefore not grounded in any definite certainty regarding the cogito, but rather in an *epistemological insecurity* (Bordo 1987). This insecurity finds its most poignant expression in Descartes' thought experiment, in which he imagines that a malicious demon exists, a 'malignant genius, whose resources and diligence are all directed towards deceiving [him]' (Descartes 1960: 110). In light of this, he asks: 'What is left that we can think of as true?' (108). In response to his question, Descartes (110) comes up with only one impregnable certainty: namely the *ego cogito ergo sum*.

> And now I have found it; for thought is the one attribute that cannot be wrenched from me. I am, I exist: that is certain. But for how long? For as long as I think. If I ceased to think, I might very well cease to be, or to exist, at that moment. So now I am admitting nothing but what is necessarily true; I am, by definition, a thinking thing [substance], that is to say, a mind or soul, an understanding or a rational being, terms of which the meaning has hitherto been unknown to me. I am a real thing, truly existent.

According to Descartes, the only certain evidence that we have to prove that we are indeed something, and not nothing, rests in our ability to think. Descartes thus eliminates his anxiety by focusing on an authentic cogito, capable of distinguishing in a rational way between different existential states (Williams 2001).

In his analysis of the subject, Descartes is concerned with developing a formal theory of the subject, one with universal validity (Cilliers and de Villiers 2000). This general assumption of universality allows Descartes to unproblematically use his own experience as a paradigm example of existence, defined as a universal, certain, and indubitable attribute of all selves. Furthermore, the emphasis on rationality enables Descartes to frame his questions independently of context. This method of analysis facilitates the development of a 'timeless, permanent structure of the self that does not change in a contingent world' (227; 228).

This understanding of the self, and the corresponding metaphysical worldview that supports it, has met with much criticism by figures such as Nietzsche, Marx,

Freud, Heidegger, and Wittgenstein. In *Beyond Good and Evil*, Nietzsche (1989) draws our attention to the fact that Descartes' certain and indubitable belief that 'I think therefore I am' (*cogito ergo sum*) is based on a 'series of daring assertions that would be difficult, perhaps impossible to prove' (23). This belief includes the assertions:

> that it is *I* who think, that there must necessarily be something that thinks, that thinking is an activity and operation on the part of the being who is thought of as a cause, that there is an "ego," and, finally, that it is already determined what is to be designated by thinking—that I *know* what thinking is (23).

What Nietzsche essentially achieves in his deconstruction of the Cartesian subject is to problematise both the ontological status of the *sum* (Derrida 1987a), and Descartes' belief in the immediate certainties of thought; a belief which is premised on a metaphysics of presence. Contrary to Descartes, Nietzsche views being in terms of a complex process of becoming, which requires the worldly domain as the arena 'in which the performance can be played out' (Cilliers et al. 2002: 5). He further argues that this process of becoming is marked by a multiplicity, which is also the generation and becoming of a will to power.[1]

Heidegger, in turn, believes that the Cartesian humanist project is misguided because—in concentrating on consciousness, spirit, and man—it fails to take into account important questions concerning the Being of beings (Howells 1998). As such, Descartes fails to provide us with an ontology of the subject. Heidegger's critique of Cartesianism is twofold: that the world is something blankly external to human consciousness, and that a subject's behaviour is merely an outward effect of his inner mental workings (Glendinning 1998). Both these assumptions are challenged by one line of thought, namely that 'the kind of Being-in-the-world that we *are* is world-disclosing' (63).

Heidegger (1962) illustrates this point by orientating his enquiry around the everydayness of *Dasein* or the way of being-in-the-world with which we are already familiar. However, as with Nietzsche, Heidegger also challenges the metaphysics of absolute presence, by arguing that being is not entirely understandable; it is also 'shrouded in darkness' (44). As explained in the previous chapter,[2] our everyday practices and our way of being in the world distract us and cloak the deeper nature of Being. Being consequently withdraws from sight. In order to recall the question of Being, we should focus our attention on the ontico-ontological difference, i.e. the difference between the concrete, everyday realities of being and the deeper, underlying structures of Being. Heidegger writes: 'Dasein is ontically "closest" to itself, while ontologically farthest away; but pre-ontologically it is surely not foreign to itself' (58). Thus, although our ontology, or the deeper structures of Being, is not as immediately apparent to us as our everyday realities, we are the type of beings who are—by nature—concerned with the nature of our being. This concern for our being manifests in the act of interpretation, understood as a questioning of '*the*

---

[1] See Sect. 3.3.

[2] See Sect. 4.2.1.

meaning of the Being of the "sum"' (53). The act of interpretation thus challenges the self-presence of Cartesian consciousness.

Despite the many philosophical critiques of Descartes' humanist subject, the Cartesian model of the self still remains highly influential to the extent that it continues to inform the commonly accepted predicates of subjectivity. These predicate, in turn, give rise to a highly specific (though widely shared) view of responsibility. The predicates ascribed to the humanist subject include a strong view of agency, intentionality, rationality, and causality, and these predicates are assumed to be *a priori* givens (see Seabright and Kurke 1997). In this view, moral responsibility is ascribed to rational, intentional, and autonomous individual agents. These agents are believed to make decisions based on reasonable principles and calculations, and the trajectory from decision to outcome is viewed in terms of a linear causality. This means that agents can be held morally accountable for their decisions, as the effects of these decisions can be determined in advance. Moreover, since the predicates of subjectivity are *a priori* and universal, all subjects are deemed to be alike. The mental life of a person is seen to 'encompasses rational calculation, intuitive ideas, intellectual deliberations and sensory inputs' (Cilliers and de Villiers 2000).

Paul Cilliers and Tanya de Villiers argue that, from the perspective of complexity thinking, the two greatest problems arising from the Cartesian view of the self is the negation of an embedded-embodied mind as a constitutive condition for human identity, and insensitivity for the ethical-political dimension of identity formation. Similar to the critics of metaphysics such as Nietzsche, Heidegger, and the post-structuralists that followed them, complexity theorists do not subscribe to a monolithic and essentialist view of the self. Rather, from this perspective, the self is seen as constituted through virtue of a dynamic process. The self emerges over time and in a network of relations with other selves and with the world. This process of identity formation introduces complexity. Furthermore, Cilliers and de Villiers argue that to treat the self as pre-programmed (in other words, to subscribe to a neutral and objective view of the self) does violence to the subject in question, because it leads to, what they term, the 'colonialisation of the subject' (242), i.e. the disregard of difference and the endorsement of social homogenisation. This is because if we presume to know who the subject in general is, we also presume to know who an individual is and how she should be treated. Furthermore, in negating the emergent and relational nature of identity formation, the ethics and politics of identity formation is also denied. Cilliers and de Villiers explain as follows:

> We are formed by social interventions (or lack of them), and others are shaped in the same way. We are not completed subjects that have to make political decisions; we come to be through those decisions. The choice to abstain from certain actions is also a political choice (242).

From this short description, it should be clear that the type of ethics that is introduced by a complexity perspective is vastly different to the traditional views on responsible and accountable action, and as such requires further investigation. However, before undertaking such an investigation, it is important to explore in more detail views on the subject that do not accord with the Cartesian humanist subject. As succinctly explained by Nancy, the critique and deconstruction of both the humanist subject and the concomitant notions of 'self presence, of consciousness, of mastery, of the individual or collective property of an essence' (Nancy 1991a: 4) has led to the so-called liquidation of the subject during the second half of the twentieth century. However, as he points out in the preface to *Who comes after the subject?,* this critique has neither led to the obliteration of the subject (an interpretation that denotes nihilism), nor to a simple '"return to the subject"' (5). Rather, there is 'a move forward toward someone—some one else in its place', and Nancy further notes that 'this last expression is obviously a mere convenience: the "place" could not be the same' (5).

I attempt to provide some insight into who this 'some one' might be, by exploring post-structural accounts of the subject. This is achieved by drawing on Levinas's understanding of the Other; Derrida's work on the subject, animals, and eating; and, Nancy's reading of Heidegger from the perspective of *Mitsein* as opposed to *Dasein*. My aim is not to provide a comprehensive overview of the post-structural re-inscription of the subject (indeed, as Nancy (3) argues, the whole debate is marked by 'a very great diversity'). Rather, I try to put forward a view of subjectivity that is not predicated on those daring assertions made by Descartes. The insights that this analysis holds will be compared to Cilliers' complex view of identity formation. The ethical implications that emerge will be explored at the hand of the example of the foreigner or what Nancy called the intruder (*L'Intrus*), as well as in terms of the ethical responsibilities that a non-humanist view of identity introduces. The traditional understanding of moral responsibility and accountability will also be critiqued, and an understanding of these concepts that is consistent with the complex view of identity formation will be forwarded.

5.2 Levinas

5.2.1 To the Other

In a chapter titled 'Difference, Identity and Complexity', Cilliers (2010a: 5) argues that the identity of systems is the outcome of differences in so far as 'relationships of difference *constitute* complex systems.' Cilliers develops this idea at the hand of Saussure's and Derrida's insights into the nature of meaning. This analysis will be presented later,[3] but at this juncture I merely wish to flag the importance of *relationships of difference* for identity formation, as this has been a central theme in continental philosophy since the time of Heidegger. As mentioned in the introduction, Heidegger frames this problematic in terms of the ontico-ontological difference between being and Being. This difference serves to decentre the worldly phenomenology of the subject, by drawing attention to the deeper structures of being.

[3] See Sect. 5.3.1.

In the analysis that follows, the various interpretations given to these relations of difference will be traced through the work of Levinas, Derrida, and Nancy. Although the focus is on developing a non-humanist, complex view of identity, I nevertheless open the discussion with the humanist philosophy of Levinas. This is because—in responding to Heidegger's view of Being—Levinas developed a highly influential philosophy of difference, premised on the recognition of the significance of otherness. In particular, Levinas had a significant impact on Derrida's thought (who, in turn, influenced Nancy's oeuvre). Levinas thus set the tone for a specific interpretation of post-structuralism, wherein the debate is focused 'under the rubric of the same and the other' (Watkin 2007: 50).

In the first chapter of Adriaan Peperzak's (1993) insightful introduction to Levinas's philosophy, titled *To the Other*, he carefully traces the influence of Heidegger's (and Husserl's) phenomenology on Levinas's thought. Levinas, a Lithuanian Jew by birth, deeply admired Heidegger's work, and was profoundly shocked by his rectoral address of 1933, in which he openly supported Nazism. Subsequent to this event, Levinas's reading of Heidegger took on a critical edge in that Levinas identifies a vulnerability to nationalism (especially Nazism) in Heidegger's work, which he sought to avoid in his own work. In *Otherwise than Being* and *Totality and Infinity*, Levinas (1998; 1969) argues that this vulnerability chiefly comes to the fore in Heidegger's view of Being. The strengths of Heidegger's project rest on the distinction that he draws between the totality of beings and Being as such, and on the fact that he does not identify *Dasein* with an autonomous subject or transcendental ego. Yet, Levinas questions the success with which Heidegger maintains both these positions.

As concerns the first point, Levinas teases out the difference between Being and beings at the hand of two meanings of the French word *être* (to be), namely '(1) that by which all beings are given as what and how they are ("Being itself"), and (2) (the whole of) reality as such, that is, (all) beings insofar as they exist' (17). Levinas argues that Heidegger conflates this difference, with the consequence that 'the beingness of Being' (i.e. the nature of Being) is nothing other than 'the beingness of beings' (i.e. the factual existence of beings) (17). Otherwise stated, it is Levinas's claim that the ontico-ontological distinction collapses, with the consequence that Being becomes 'the monoistic horizon or source of universal participation' (17). Secondly, although Levinas concedes that Heidegger radically transforms the philosophy of consciousness through his analysis of Dasein; he nevertheless argues that 'the central and all-mastering position of the modern ego is also retained in Heidegger work' (17). This is true to the extent that Heidegger characterises other beings 'only as companions within anonymous communities' (17), rather than as beings that can decentre the ego.

Levinas further maintains that in Heidegger's characterisation, beings, in their common orientation towards Being, are always already constituted in terms of a 'prepredicative familiarity' (18). According to Levinas, Heidegger therefore gives a positive interpretation of our thrownness into the world, signified by the expression *es gibt* (there is). For him, the *es gibt* is the condition by which Being sheds light onto being. Levinas however is unconvinced that the *es gibt* (or what he translates as the *il y a*), constitutes such a generous expression. Rather, he views the *il y a* as characterised by indeterminate forces that threaten my subjecthood.

Levinas (1988) introduces his phenomenology in a small post-war publication titled *Existence to Existents*. In this work, reality is defined as a plurality, rather than as a monism wherein everything unfolds around one basic instance of '"the Same"'[4] (19). In the preface to this publication, Levinas (20) describes the threat and the fear of the anonymous current of the *il y a* as follows:

> It is because the *there is* has such a complete hold on us that we cannot take nothingness and death lightly, and we tremble before them. The fear of nothingness is but the measure of our involvement in Being. Existence of itself harbors something tragic which is not only there because of its finitude. Something that death cannot resolve.

The tragedy of existence concerns the impersonal and anonymous nature of being. Yet, in this publication Levinas explains how the journey towards truth constitutes an escape from anonymity. It is thus a journey away from existence to existents (i.e. to subjects or egos). Peperzak (1993) notes that this journey (which is also captured in the title of Levinas's publication) constitutes a reversal of Heidegger's thought. He explains as follows:

> Whereas the latter started from a reflection on beings (*Seinendes, l'étant,* or *l'existant*) in order to discover Being (*das Sein, l'essence,* or *l'existence*), Levinas described the way of truth as a movement from "essence" or "existence" (*Sein*) to "existents" (*Seinendes*) (5).

Levinas (1988) argues that the ontological journey from existence to existents takes place via a *hypostasis*, which is that event whereby a subject emerges from anonymous being (the *il y a*) and arises through consciousness in order to exercise mastery over being. The present is the event of the *hypostasis*, and as such it tears into the infinite series of undifferentiated time. The present is however always already fading away. Being, which occurs in the instant, is thus also characterised by evanescence. The nature of being therefore constitutes a continual process of renewal or rebirth in which we affirm our subjectivity in the next instant, but it also implies effort and fatigue as we struggle for our future.

Having escaped the anonymity of existence-in-general we find ourselves in a solitude created by the definitiveness of the bond with which the ego is chained to itself. Levinas's statement that '[t]he world… [is] solitude' (84) illustrates the solipsistic manner in which we engage with the world. In this regard, Levinas affirms Hegel's view that each of us is the centre of our world. What this implies is that although we engage freely in the world, and although we can distance ourselves from the objects in the world, we can never be free of ourselves. As such, we are always alone. As Levinas states: 'In the understood universe I am alone, that is

[4] Interestingly, Descrates also endorses the view that the self is not centred on one basic instance of 'the Same'. In 'The third meditation', Descartes argues that human consciousness does not only contain an idea of itself, but all the irreducible ideas of the infinite. Consciousness thus relates the self with a reality that remains irreducible to the self. Although Descartes identifies the infinite with God, Levinas nevertheless uses the basic structure of Descrates's argument to put forward a characterisation of the Other as that which manifests beyond consciousness (and that cannot therefore be understood from the perspective of consciousness) (Peperzak 1993: 21).

closed up in an existence that is definitively *one*' (85). This is not to say that we do not encounter or perceive others (indeed, the fact of another facing me is proof of this), yet we do not *experience* otherness. This is because the other's otherness is immediately assimilated my self-centred consciousness. My experience of the other thus constitutes a phenomenology of *egology* (Peperzak 1993: 19), in that I inhabit myself entirely.

Yet, there is a second form of engagement with the other, in which the otherness of the Other is experienced. For Levinas (1988), this meeting constitutes the breaking-up of the categories of the ego, since I find myself confronted by someone other than myself. In the confrontation with the Other, I am exposed to the alterity of this Other. This radical alterity however cannot be rendered intelligible or understandable by the categories of reason. Therefore, on the one hand, the presence of the Other makes me aware of my own solitude, makes me realise that self-escape is impossible, and that I am alone. But, on the other hand, the alterity of the Other confronts me with something wholly different to myself, it allows me to recognise something that I am not and that my existence is not. The alterity of the Other (unlike the alterity of death or of the future) is an otherness that neither destroys my subjectivity, but nor can it be assimilated to become part of my being. The Other is thus other than being; the Other transcends the limits of consciousness, and signifies a beyond to phenomenology. This is because the Other cannot be grasped and thus remains incomprehensible to thought. Unlike Heidegger's phenomenology, the Other's otherness thus escapes totalisation in that the Other 'comes towards me as a total stranger and from a dimension that surpasses me' (Peperzak 1993: 20).

This experience of otherness 'opens from the outside in the face of another, in the other who faces' (Lingis 1988: 11). The nakedness of the Other's face, and the experience of the Other's irreducibility, simultaneously threaten my ego and compel me to accept responsibility for the Other. Christina Howells (1998: 125) explains as follows:

> the otherness of the Other as free transcendent subject both arouses my hostility and is also what causes it to cease, in so far as the face initiates an experience of transcendence and freedom which commands respect for the Other. And this respect is the primary imperative: the 'incarnation of non-violence'.

In the confrontation with the Other, the imperative to not kill the Other comes to the fore and 'I "know" myself to be *obligated*' (Peperzak 1993: 22). Yet, this obligation is felt pre-theoretically to the extent that the fact-value distinction—'the scission between factuality (*be*) and normativity (*ought*)' (22)—has not yet emerged. Simon Critchley (1999a: 3) refers to this encounter as 'the primordial ethical experience', which serves to ground both substantive ethics and theoretical philosophical insights. This primordial ethical experience carries the consequence that, henceforth, '[o]ur existence will no longer appear… as destined for the world' but rather for 'being-responsible for the other human' (10–11). Derrida (1995: 279) contends that this is a mode of subjectivity 'constituted first of all as the subjectivity of the *hostage*'. The ethical obligation towards the Other means that we no longer live for the satisfaction of our selfish desires, for the reason that '[t]he subject is responsible for the other before being responsible for himself as "me."' (279).

The meeting with the Other however not only destines the self-for-the-Other, it also redeems the self. This is because, in challenging or questioning my primacy as subject, the face of the Other frees me from reification, pulls me from the senseless rumblings of my existence, and grounds the ethical. This redemptive alterity granted by the Other defines the ethical relation. In *Totality and Infinity,* Levinas (1969: 43) describes ethics as 'the putting into question of my spontaneity by the presence of the other.' In other words, ethics concerns an irreducible otherness. It poses a challenge to the liberty and spontaneity of the ego by virtue of tying this ego to a point of otherness or exteriority.

5.2.2 The Problem of Asymmetry

Despite his powerful and highly original analysis of the encounter with the Other, Levinas's conception of the asymmetrical relation with the Other nevertheless also introduces a number of problems. The first problem concerns the fact that he provides us with a view of a divided humanity. Ernst Wolff (2010: 138–139) describes this problem as follows:

> [humanity] doesn't mean the same for other and for the self. For the *other* "being human" means to exert a pre-predicative imperative or questioning on the self... for the *self* "being human" means to have one's identity before anything else in the assignment to respond to the other.

Related to this criticism is the fact that, since every 'I' is another's 'Other', all humans are equally obligated, meaning that the ethical experience becomes a universally-shared (if not banal) experience. Levinas (1998) attempts to address this critique in *Otherwise than Being*. Here he argues that the original asymmetry between self and Other should not be obscured by the secondary truth of our equality, whereby each person is 'a replaceable instance of one universal "being human"' (Peperzak 1993: 28). In order to safeguard this original asymmetry, Levinas distinguishes between the *Saying* (*le dire*) and the *Said* (*le dit*). Whereas the former refers to an irreducible exposure to the Other, the latter concerns intelligible meaning as that which is transferred through communication. Levinas (in Levinas and Kearney 1986: 29) expresses the difference between the *Saying* and the *Said* as follows: 'Language as saying is an ethical openness to the other; as that which is said—reduced to a fixed identity or synchronized presence—it is an ontological closure of the other.' The primordial experience that manifests in the *Saying* therefore cannot be thematised and cannot correspond with the *Said*. Peperzak (1993: 29) notes that '[w]hereas the *Saying* breaks all the limits of philosophical language, the *Said* belongs to the dimension of things that are objectifiable.'

Drawing from the terminology introduced in Chap. 3, we can argue that Levinas safeguards the asymmetrical relation with the Other by framing the experience in terms of an aneconomic force, which can never be translated into utilitarian, economic forces. In this interpretation, asymmetry refers to the event of the Other, which does not necessarily correspond with the Other as person, but rather denotes an irreducible openness to otherness as such. Like Bataille, Levinas thus affirms an absolute distinction between the aneconomic and the economic realm. Indeed, Levinas argues that although our needs are satisfied within the economy of the ego (which is made up of our corporeal and terrestrial experiences), true desire (as fulfilled in the encounter with the Other/otherness) points beyond the horizon of economy (Peperzak 1993).

The absolute asymmetrical relation between the self and the Other however poses a second difficulty for Levinas: the Other's otherness is so other that no relation between self and Other is, in fact, possible. To understand why this is the case, it is useful to follow Cilliers in his analysis of difference. Cilliers (2010a) writes that in order to recognise a difference between A and B, three conditions must be in place: firstly, we must be able to identify them as A and as B, secondly, there has to be a common factor (i.e. some element of similar identity) between A and B so that we can relate them to each other, and thirdly (but related to the second point), A and B must be situated in a larger network of differential relations so as to allow for meaningful comparison. Otherwise put, '[o]ne cannot talk of the differences between A and B if they are the only two things under consideration' (12). The implication of this analysis is that in order to recognise the other as other, the other and the self need to share some form of identity. The asymmetrical relation between self and other, in other words, cannot be absolute. Cilliers (2010a: 12) explains as follows:

> If... the notion of difference is absolutised, it may lead one to think that *no* relationship between the self and the other is possible; that the other is absolutely other. However in order to be able to recognise the other as other at all, some form of identity between the self and the other is required. As a matter of fact, the claim that the other is completely unknowable is nothing but an inverted insistence on *pure* identity...

As is the case with pure identity, absolute alterity is impossible, which, paradoxically, seems to threaten the very otherness of the Other that makes out the basis of Levinas's ethics. This is because, as Badiou (2001: 24) notes, 'the celebrated "other" is accepted only if he is a good other—which is to say what, exactly, if not *the same as us*?'. In other words, in practice, talk of the wholly Other tends to slide back into talk of the same, since we are incapable of recognising, or of relating with, absolute alterity.

In his essay on Levinas titled 'Violence and metaphysics', Derrida (1978: 111) recognises the logical impossibility of Levinas's project in calling his views 'an Ethics without law and without concept'. Yet, he contends that this is 'not an objection', since Levinas 'does not seek to determine a morality, but rather the essence of the ethical relation in general', and that this is achieved by placing the possibility of ethics in its impossibility. According to Robert Bernasconi (1987: 135), this interpretation 'preserve[s] the *thought* of the ethical relation (a thought which is not yet a practice) rather than the ethical relation itself.'

Preserving Levinas's impossible ethical relation holds the risk that 'the very possibility of something singular would remain irretrievably lost' (Gasché 1994: 13). For this reason, Derrida argues that one should not prioritise Levinas's ethical project (which is concerned with infinite responsibility) over Heidegger's ontological project (which is concerned with the phenomenology of being). Rather, ethical asymmetry (the respect for the Other's otherness) must operate within phenomenological symmetry (i.e. the recognition of the other as ego) (Cornell 1992: 85), because '[i]f the other was not recognized as ego, its entire alterity would collapse' (Bernasconi 1987: 125).

In conclusion, the strength of Levinas's project lies in presenting us with a non-totalising view of existents. Non-totalisation is the outcome of the asymmetrical relation between the self and the Other. It is through means of the subject's relation to the Other that the solipsistic ego becomes decentred, and that her identity is reconfigured in light of the imperative to respond to the Other. Yet, the problem of a divided humanity (in which the subjectivity of self and Other remain irreconcilable) does not provide us with a convincing account of identity. The central problem hinges on an asymmetrical view of relationality, since this does not allow for a meaningful relation between self and Other. This is because, in endorsing this asymmetrical view of relationality, Levinas denies the kernel of a common identity that is needed to facilitate relationality and recognition. On the final count, Levinas thus provides us with an essentially unworkable account of both the Other and of ethics.

Levinas's view of ethics is further problematised by the fact that, in practice, absolute responsibility for the Other is always destroyed by the face of the third—that is, another Other that make equal, yet unique, demands on the self. In practice, the ethical relation therefore always already gives way to politics, in which utilitarian calculations override untenable ethical obligations. The stark contrast between ethics and politics, or the aneconomic and the economic, ultimately means that the *Saying* is silenced in a world dominated by the *Said*. The question that now remains is whether deconstruction—which is also a philosophy of difference—can enact, as opposed to merely think, the ethical relation.

5.3 Derrida

5.3.1 The Question of the Animal

Like Levinas, Derrida is also concerned with preserving the alterity of the other. Indeed, it will be shown that the question of identity, or what Derrida (1995) calls the question of the *who*, implies an even more radical alterity for Derrida than is the case for Levinas. This is because the question of the *who* concerns that which is no longer exclusively associated with the subject. However, unlike Levinas, Derrida is also sensitive to phenomenological questions that arise from our being together in the world. This double sensitivity for alterity and mutual dependence comes strongly to the fore in an interview with Nancy, titled '"Eating Well," or the Calculation of the Subject' (1995), as well as in his book titled, *The Animal that Therefore I Am*

(2009). The analysis will proceed at the hand of the interview, but before turning to this work, it is firstly necessary to provide an overview of Derrida's view of difference, which was deeply influenced by Saussure's (1960) view of language.

Saussure demonstrates that there is no original core to signification: all meaning is already mediated through signifiers and properties that are differentiated from one another. For example, we recognise the colour orange, not because of its innate orangeness, but because of the differences between orange and other colours such as green, purple, or blue (and between colours and other properties as such). Meaning therefore resides in the relations between signifiers or properties, and not in some intrinsic characteristic of a given signifier or property. Following Saussure, Derrida argues that there is no exact or literal meaning that is inherent to language—rather, all we have is a system of differences, where every event or every speech act is itself made possible by prior structures (Culler 1983).

These structures concern not only how the network is arranged, but also the distinctions or *constrained differences* that serve to define the terms in the network as such. Cilliers (2010a: 10) explains that '[t]he meaning of a component at a specific point in the history of the system is therefore that which satisfies all the current constraints placed on it through all its relationships in the current context, i.e. as determined by the current boundary.' This means that random differences are not enough for establishing identity. In Derrida's terminology, differences only count as differences when they are related to the logos, and hence fall within the conceptual schema or hierarchy that supports the logos. Identity or meaning is thus the product of *constrained difference* (11).

Cilliers also notes that differences should not be understood in terms of binary oppositions, but in terms of differential relationships (to say that A differs from B is not to say that B is not-A). This differential, as opposed to oppositional, understanding of language informs Derrida's (1976) view of meaning, especially as articulated in *Of Grammatology*. Cilliers (2010a) further argues that this differential understanding of meaning is only possible if components are richly interconnected. To reiterate: complex identity cannot be thought of in terms of only two components (an 'I' and a singular other), but must be emplaced in a network of relations. Given this differential account of meaning, the face of the third that serves to problematise Levinas's analysis is thus critical for conceiving of a complex identity.[5]

The above account of meaning affirms Saussure's insight that the movement of signification takes place in a *regulated* economy of passive spacing and active deferral from one unit of meaning to the next, and that this movement gives rise to

[5] Cilliers (2010b: 59) writes that boundary formation is the process by which possibility is 'actualised in the presence of constraints'. Cilliers also notes that the '[t]he fewer [the] constraints, the more possibility, but possibility left empty' (59). For example, a breadth of possibility is open to a small child who has an entire lifetime ahead of her and who is still relatively unconstrained by life choices. Conversely, '[t]he more constraints, the better we can get at meaning, but the more bountiful it is' (59). For example, as one gets older—and exercises more choices in terms of career, spouses, offspring etc.—the many possibilities of childhood begin to close down, but there is a richness and depth to one's life experiences (exercised within very definite constraints) that is missing from the life of a child.

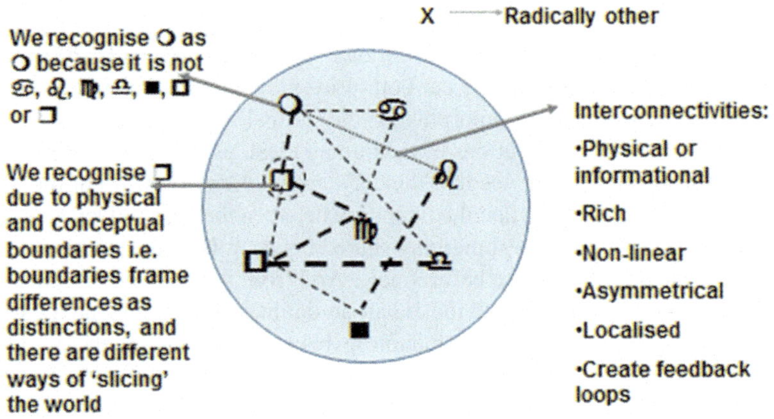

Fig. 5.1 'The relational nature of identity'

identity (see Fig. 5.1.). However, Derrida also criticises Saussure for viewing language as a relatively stable system of signs that can yield predictable responses. Rather, and as explained by Cilliers (14), 'identity is the result of diversity, and if differences are constantly being moved around in feedback loops and imposed from outside the system as the context changes, then identity is by definition a dynamic concept.' Derrida (1982) argues that the movement of signification can neither be arrested nor be described analytically because it is defined by the workings of *différance*, which consists of both an economic dimension (the Saussurian regulated economy of difference) and an aneconomic dimension. The aneconomic dimension characterises meaning as excessive and without origin, and hence, as an expenditure without reserve.[6] Essentially, the difference between these dimensions boils down to that which we can express in the language of concepts and that which we cannot. As such, it concerns an understanding of *différance* as that which generates a relatively stable identity (as per the view of meaning pioneered by Saussure), and that which ruins or destabilises identity (and that may even destroy identity).

Derrida tackles the implications that this dual view of *différance* holds for identity at the hand of the question of the animal. For Derrida (in Derrida and Roudinesco 2004: 63), the question of the animal represents 'the limit upon which all great questions are formed and determined, as well as all the concepts that attempt to delimit what is "proper to man," the essence and future of humanity, ethics, politics, law, "human rights," "crimes against humanity," "genocide," etc.' The two readings of the animal that the *aporia* of *différance* gives rise to are discussed below.

On the one hand, recognising the animal means recognising the singularity of the animal. This means that I am capable of distinguishing a particular animal (Derrida uses the example of his cat) from other animals, and from all other subjects and

[6] See Sect. 3.4.1.

objects in the world. In this regard, Derrida writes of the singularity of a particular cat that it is not 'the exemplar of a species called cats' (Derrida 2002a: 378). Recognising the singularity of the cat both prevents me from framing the cat as a general other, and of anthropomorphising the animal and casting judgements of animals 'as only partial realizations of the human ideal, as subhuman' (Wood 1999: 20). Derrida (2002a: 381) writes that the cat's gaze 'offers to my sight the abyssal limit of the human', meaning that the boundary between the cat and Derrida remains firmly intact. Yet, there is also something shared between Derrida and his cat, since, as he notes, 'the animal is there before me… And from the vantage of being-there-before-me it can allow itself to be looked at, no doubt, but also… it can look at me' (380). In recognising a shared phenomenology between himself and the cat, Derrida is able to avoid the criticism that Badiou levels against Levinas, namely that the celebrated Other slides back into talk of the same, because we cannot relate with absolute alterity (which is another way of saying that we cannot relate with that which we cannot recognise as different to us).

Yet, if I am able to look at the cat and it at me, then this experience raises the question of whether the singular animal is capable of responding to me in the face-to-face encounter. Derrida engages with the idea that the animal is, in principle, capable of responding, understood here as 'answering to' or being responsible for a singular or proper name (Spivak 1994). If this is the case, one cannot pose 'a single, linear, indivisible, oppositional limit, to a binary opposition between the human and the infra-human' (Derrida 1995: 285). Rather, we need to contend with the rights and obligations that are always conditional and conditioned by the animal's singularity, the animal's name (Derrida 2005). In other words, we must extend the law of hospitality (introduced in Sect. 4.5.2) to the animal other.

On the other hand, questions concerning the animal and our responsibilities to the animal are excessive and incalculable, and do not allow for delineation or assimilation. Under this reading, the question of the animal undermines both 'the phrase "the animal" and how it is used oppositionally to define "the human"' (Calarco 2004: 191). The question of the animal necessitates that we take seriously questions such as whether the animal can hear the call to responsibility, whether the animal can question, whether the call heard by *Dasein* comes originally from the animal, whether there is an advent of the animal, and whether the voice of a friend can be that of an animal (Derrida 1995). In raising the question of the animal, Derrida again seeks to contest the existence of an oppositional limit. However, in challenging the oppositional limit, the question of the animal also becomes the question of the human. This is because if human privileges can no longer be defined or justified by referring to the category of the animal-in-general, then this begs the question 'Who is the human?'. Practically speaking, this amounts to the fact 'that we never know, and never have never known, how to *cut up* a subject'[7] (285).

[7]This concession not only necessitates that we take into account discriminatory attitudes towards animal societies, but also, as Derrida (1987b: 183) argues, partitions and separations 'other than Auschwitz—apartheid, racial segregation—other segregations within our Western democratic

When speaking about the subject, we need to contend with the undecidability that resides at the heart of the notion of the subject. In order to progress responsibly when faced with the unknown, we have to heed to the responsibility of justice, which is also 'the unconditional law of unlimited hospitality' (Derrida 2005: 6). Justice demands that we stay open to, and prepare for, the before and the after of the subject who/which precedes classification. In terms of the animal, this arm of the *aporia* necessarily evokes a certain incalculability that prohibits us from defining or categorising the animal in advance, and of assigning her a proper name. To try to determine what or who the animal is, or how one should treat the animal, invariably implies violence to the being in question. In order to try to mitigate this violence, our notions of both the animal and our responsibilities to the animal must pass 'through the proofs of the incalculable and undecidable' (Derrida 2002b: 273). To fail to do so is to reduce everything to calculation, and hence to generally determined categories of being. Consequently, Matthew Calarco (2004: 191) argues that if 'Derrida has accomplished anything in posing the question of the animal, it has been to raise these very limitations as questions.'

5.3.2 Deconstructing Humanism

Although Derrida applauds the work done by Hegel, Nietzsche, Heidegger, and Levinas in disrupting the traditional view of humanism; he also takes issue with these thinkers because they never undertake the radical questioning of the *who* that Derrida proposes. Despite their respective displacements of the subject, they continue to endorse the *who* as subject (and moreover a *human* subject), and for this reason Derrida argues that they remain profound humanists.

Focusing specifically on Heidegger, Derrida (1995) argues that it is ultimately around the privileging of the question of Being—questioning in turn being a *human* privilege—that 'the metaphysics of subjectivity' (262) was constituted. In other words, Heidegger identifies the *who* as already being there as the force or power to ask questions (Derrida 1995). Although exceptions can be found, most Heideggerian and post-Heideggerian discourses continue to privilege the human on the very basis of the unquestioned *who*.

Heidegger's humanism is also evident in his view of animals' relations with the world. Derrida notes that, for Heidegger, the animal has no world, nor any environment (*'Das Tier hat keine Welt, auch keine Umwelt'*) meaning that '[a]nimality is not of spirit' and the animal is therefore *weltarm* or poor in the world (Derrida 1987a: 47). This negative formulation of 'not-having' points to the Heideggerian idea that '[t]here is no category of original existence for the animal' (Derrida 1995: 277). Otherwise put, the animal 'has access to entities but, and this is what distinguishes it from man, it has no access to entities as such' (Derrida

society. All these differences have to be taken into account in a new fashion.' This is because repressing these differences denies an engagement with the problematic of the subject.

1987a: 51). Derrida explains that this is because, for the animal, the Being of the entity is crossed out in advance. This is an absolute crossing-out, which signifies privation. Consequently, *Dasein's* definitive predicate 'is always a matter of marking an absolute limit' (54) between *Dasein* and the animal, where 'the animal is absolutely off-limits to the deconstruction of ontology' (Spivak 1994: 31).

In the interview, '"Eating Well"' Derrida (1995) embarks on the task of deconstructing the limits of the subject in order to open up the notion of the subject and displace the related ethical frontier. Derrida argues that traditionally the ethical limit between subject and non-subject is marked by the injunction to non-violence. Responsibility for the other is expressed in Levinas's imperative, 'Thou shalt not kill'. However this mandate, which addresses our relation with the other, presupposes the other as human other. As such, the 'Thou shalt not kill', and the related primordial ethical experience, has always been understood as 'Thou shalt not kill man' and never as 'Thou shalt not kill the living in general'.[8] For Derrida (2003: 113), Heidegger's and Levinas's humanism is thus characterised by '*the extent that they do not sacrifice sacrifice.*'

In highlighting this ethical limit, Derrida (1995) is attempting to underscore the canonised sacrificial structure that marks the Western metaphysical discourse. This sacrificial structure is characterised by the acts of 'ingestion, incorporation, or introjection of the corpse' (278)—an operation which is symbolic when the corpse is human; but real, as well as symbolic, when the corpse is an animal. According to Derrida, it is this notion of sacrifice which is 'essential to the structure of subjectivity' (Derrida 2002c: 247), in that it creates an absolute limit between the subject and sacrifice.

This limit does not recognise the radical irreducibility of the singular Other. Rather, the limit that marks sacrifice is expansive and general insofar as 'those singular beings that we call "human" are named and configured *as* human' (Calarco 2004: 193). Otherwise stated, symbolic sacrifice universalises 'a singular being under the name "human"' (193). Similarly, the act of flesh eating, the 'consecration of flesh-sharing *is* its erasure, the spiritualization and denegation of its gory reality' (Clark 2004: 103). Thus, partaking collectively in the flesh of the singular animal enables us to paradoxically distance ourselves from the animal, symbolise the animal, and reduce its singularity. This opens up the space for a very real non-criminal putting to death of the animal, which is an act that can be justified on the grounds of symbolic sacrifice and by postulating the only 'real' entity as the human subject (Derrida 1995).

Note that Derrida's point is not that real and symbolic sacrifice should be equated, nor that sacrifice should be done away with—indeed, as will be shown—sacrifice is inescapable. Rather, Derrida calls our attention to the notion of sacrifice in order to show how it has, until now, constituted both the material and symbolic outcome of the entire anthropocentric tradition without ever having been sufficiently deconstructed.

---

[8] Indeed, Calarco (2004: 181) notes that '[b]esides the short essay in *Difficult Freedom* entitled "The Name of a Dog, or Natural Rights," Levinas appears wholly uninterested concerning the relation of animals to the ethical or justice-as-politics.'

This tradition is defined by what Derrida (280) calls the *carno-phallogocentric structure of subjectivity*, in which the virile figure of the self-present, speaking, male eater of flesh[9] is installed at the determinative center of the subject (Derrida 1995; Calarco 2004). In this schema, strength and virility do not merely represent the human ability to master and possess nature actively, but are also reinforced by the acceptance of sacrifice and the eating of flesh (Derrida 1995). In other words, '[w]e have to eat and we have to eat something living. That is the law of flesh' (Clark 2004: 52). This is related to what Carol Adams (1990: 26) refers to, in another context, as 'the sexual politics of meat', where '[m]eat eating measures individual and societal virility.'

It is the consecration of flesh-eating and of sacrifice (both real and symbolic), and of the 'carnivorous' assimilation of the other, that allow for this schema of subjectivity to survive unchallenged (Derrida 1995). It is for this reason that Derrida states that it suffices to take seriously not only 'the idealising interiorisation of the phallus'; but *also* 'the necessity of its passage through the mouth, whether it's a matter of words or things, of sentences, of daily bread or wine, of the tongue, the lips, or the breast of the other' (280).

Once we start considering the ingestion and interiorisation of the singular other, the opposition between subject and sacrifice is muddied. The human condition is one of dependence: we are constantly giving of ourselves to specific others, and taking from specific others. The interiorisation of the other is thus both expansive and unavoidable, and cannot be limited to flesh-eating. As such, the threshold or line that demarcates the realm of the subject/the *unsacrificable* from the animal/the *sacrificable* (as typified in the injunction to not kill human beings) should not be viewed in terms of a simple binary opposition: a clear-cut inclusion/exclusion or inside/outside of the subject. Rather, the limit passes between 'several infinitely different modes of conception-appropriation-assimilation of the other' (281).

By uncovering the basis of the carno-phallogocentric schema, Derrida deconstructs the limit that demarcates 'the human *in general* [from] the animal *in general*' (Derrida 2003: 128), and focuses our attention on the '(symbolic or real) "eat-speak-interiorise" of the singular other'. In deconstructing the ethical limit, Derrida further shows how the singularity and irreducibility of the other, as well as the relation to self, 'requires a thinking of *différance* and not of opposition' (Derrida 1995: 269). In other words, an ethical frontier defined not as an oppositional limit but as a limit mediated by *différance,* 'insists on multiplication and complication where essentialist gestures have homogenized, reduced, or screened out important differences' (Calarco 2004: 24).

---

[9]To illustrate this point, Derrida (1995: 281) uses the example of the *chef d'Etat* (head of State) as embodying the pinnacle of this fraternal structure, in that the *chef d'Etat* 'must be an eater of flesh'. In a footnote, Derrida points to one exception to the rule, namely Hitler, who was in fact a vegetarian, but states that '[e]ven he did not propose his vegetarianism as an example' (475). Moreover, Derrida views this exception as illustrative of the 'hypostudy' which he is trying to evoke, namely: 'A certain reactive and compulsive vegetarianism is always inscribed, in the name of denegation, inversion, or repression, in the history of cannibalism' (475).

5.3.3 A New Limit: Eating the Good

Up until this point, the notion of *the Good* (which has informed our ethical, judicial and political foundations) has always been related to questions sealed within the human subject (Derrida 1995). Displacing this limit has serious implications for our conceptions of the Good, since the symbolic and real operation of ingestion must now be calculated into our conceptions of what concerns the Good. Ingestion is not limited to eating, but includes all the ways in which we give to and take from others via our sensory apparatus. We ingest the other not only through taste, but also through sight, smell, sound, and touch. These modes of assimilation have largely been ignored, because we have generalised sacrifice. In re-conceptualising the ethical limit, Derrida forces us to acknowledge, and reckon with, the many ways in which we help constitute, and are constituted, by others. Speaking of this new limit, Derrida (281) writes that 'everything that happens at the edge of the orifices (of orality, but also of the ear, the eye—and all the "senses" in general) the metonymy of "eating well" [bien manger] would always be the rule (of every morality)'. The task lies in 'determining the best, most respectful, most grateful, and also most giving way of relating to the other and of relating the other to the self' (281–282).

Derrida asserts that this task is problematised by the fact that the symbolic operation of ingestion is impossible to delimit, because the question of the *who* remains undecided and undecidable. Yet, we must undertake this task because the experience of *eat-speak-interiorise* is unavoidable, both in terms of physical nourishment (eating, suckling) and of our being-together-with-the-other in the world (in the sense of seeing, hearing, touching, and speaking to the other). Acknowledging this need for the other, and taking responsibility for *eating-speaking-interiorising* the other, undermines the sacrificial structure implicit in the history of metaphysics. Once this denegation of the real or symbolic assimilation of the other is dismantled, the violent institution of the *who* as subject becomes undone. Derrida contends that this marks the closure of metaphysics, and opens up the space for a new and bolder economy of need and desire (in which we ethically reckon with our dependence on one another).

Derrida writes that in this new economy, '[t]he question is no longer one of knowing if it is "good" to eat the other or if the other is "good" to eat, nor of knowing which other. One eats him regardless and lets oneself be eaten by him' (282). Eating and ingestion thus become metaphors for assimilation and being-together in the world. Eating is unavoidable because it is the means by which we express our desire and need for presence, pleasure and the *Da* of *Da-sein* (Krell 1988). However, Derrida (1995) contends that this economy should also be regulated by the ethical mandate, *'Il faut bien manger'*, which should be understood both as 'one must eat well' (475) in the sense of '*learning* and *giving* to eat, learning-to-give-the-other-to-eat' (282); as well as 'everybody has to eat'. The adverb *bien* should further be nominalised into *Le Bien*, to imply 'the eating of the Good' (475). Furthermore, because one never eats entirely on one's own, this ethical mandate is the rule offering infinite hospitality.

The ethics of eating (and of showing hospitality) is however frustrated by the fact that the other can never be wholly assimilated. This too must be contended with when attempting to 'eat well'. In this regard, the metonymy of 'eating well' is a response to the *aporia* of the animal. Eating well means, on the one hand, that we must be open to, and prepare for, the before and the after of the subject that/who precedes language (which can also be read as the subject that/who precedes classification). On the other hand, we must reckon with the singularity of the other—'identify with the other, who is to be assimilated, interiorised, understood ideally (something one can never do absolutely without *addressing oneself to the other* and without absolutely limiting understanding itself, the identifying appropriation)' (283). Derrida contends that '[t]he 'sublime refinement' involved in this respect for the other is always a way of "Eating well," in the sense of "good eating" but also "eating the good" [*le Bien Manger*]' (283). He also writes that '[t]he good, must be eaten and eaten well' (283). In this vein, Derrida tries to bring together Levinas's ethical relation and Heidegger's phenomenology. Enacting the ethical relation, instead of just thinking on the ethical relation (as is the case with Levinas), implies a continuous engagement with this *aporia*. In this regard, we are tasked with answering the following difficult questions:

> *how* for goodness' sake should one *eat well* [*bien* manger]? And what does this imply? What is eating? How is the metonymy of introjection to be regulated? And in what respect does the formulation of these questions in language give us still more food for thought? In what respect is the question, if you will, still carnivorous? (281).

Yet, the above questions will never receive a final answer. This is because implicit in the ethical mandate of *Il faut bien manger,* and the corresponding economy of need and desire, is the logic of *différance* that requires that the modalities and contents of this mandate be varied, ad infinitum. Therefore, whereas Levinas preserves the thought of the ethical relation in the *Saying* that cannot be *Said*, Derrida asks of us to continually and wordlessly enact the *Saying* in the world (Cornell 1992). This is achieved by constantly contemplating and responding to the ethical mandate of *il faut bien manger*.

Derrida's preoccupation with, and respect for, difference forces this difficult outcome. We are infinitely responsible for the consequences of our decisions, even though our decisions are structured by the '*experience and experiment of the undecidable*' (Derrida 1988: 116), which is to say, structured by a grappling with alterity. Unlike in traditional accounts, responsibility is no longer a choice for a Levinasian or Derridean subject. For Levinas, the face of the Other compels us to accept responsibility; whereas for Derrida our shared being-together-in-the-world means that we invariably assimilate the other, towards whom we nevertheless owe infinite responsibility and hospitality even though (or, perhaps, because of the fact that) we cannot categorically define this other.

For both Levinas and Derrida, it is the individual subject who bears responsibility towards the other. Nancy however shifts the debate from a philosophy of alterity (which is dominated by the self-other problematic) towards a view of self-as-other or self as *singular-plural*, where being implies being-with, and plurality characterises every singularity. For Nancy, the self is ontologically (and not only relationally) differentiated in its own self. Yet, since all beings share this ontological property, it is precisely the self as always-already-with-others that we share with one another. Nancy thus uses this ontological difference in order to develop a philosophy premised on relationality, as opposed to alterity.

5.4 Nancy

5.4.1 *Nancy's Reading of Heidegger's* Mitsein

Nancy is interested in presence and meaning, yet he defines these concepts as non-essentialist. Emma Campbell (2003: 43) notes that Nancy understands 'a "birth" to "presence"... not in terms of being or identity (what presence might *be*) but in terms of an opening or coming *to* presence (that coming into that lack of form or identity that characterises a pure opening as such)'. Therefore, in Nancy's work, the question of presence (coming into being) precedes and informs the question of identity (or what it means to be). Furthermore, presence is defined as a co-presence, since the coming into being is exactly what beings share.

At first glance, this understanding of being as the act of presencing seems remarkably similar to Heidegger's view. To recall from the introduction to this chapter, Heidegger abandons the view of a single, undivided, and free subject (who only relates to the community in a secondary fashion) in favour of a view of being as thrown or inscribed in the world. Ignaas Devisch (2000: 242) explains as follows:

> The ontological fact that I am already a social being prevents me from dreaming of myself as an independent being who is free to enter into sociality and who gives sense to his own life.

Indeed, in as much as being is thrown into a shared world, being is always a being-with-others. Heidegger (1962) affirms this in §26 of *Being and Time*, where he argues that *Mitsein* (being-with) and *Mitdasein* (being-there-with) are co-essential to *Dasein.* This is because, as Devisch (2000: 242) explains, '[t]he 'there' (*da*) makes of me at the same time a 'with' (*mit*). Or more exactly: the 'there' is always already a 'with''. Yet, what is striking is that Heidegger's discussion on *Mitsein* (which proceeds at the hand of an analysis of 'taking care of' as the relational mode of the *with*) only occurs after an extensive discussion on the originality of *Dasein.* This begs the question as to whether *Mitsein* and *Dasein* are truly co-existential and co-original, since the *Mitsein* does not seem to alter the ontology of the *Dasein*.

It is the problematic of the *Mitsein* (the *with*) that Nancy takes up in his own work. He argues that although Heidegger opened up the space for thinking through 'an analytic of the *coexistential*' (244), he later erases the possibility of seriously contemplating the coexistential. This erasure is due to the fact that he subordinates the *Mitsein* to a thinking of the proper and the improper, with the consequence that the *with* is lost because the '*[m]it-* supervenes... upon an already existent *Dasein*' (Watkin 2007: 55). Nancy identifies the loss of the *with* as the primary weakness in

Heidegger's conception of being. As such, he offers a re-reading of Heidegger's ontology based on an essential, yet non-essentialist understanding of the *with*, thereby recasting Heidegger's ontology as a 'social ontology' (Critchley 1999b: 240), based on the primacy of community.

Nancy (2008: 4) begins his analysis by identifying the three ways in which Heidegger conceives of *Mitdasein* in *Being and Time*, namely:

> the banal being-alongside (a common mode in the sense of ordinary, vulgar), the common as the sharing of properties (relations, intersections, mixtures), or, lastly, the common as ownmost structure in itself, and thus as communional or collective.

The first interpretation implies absolute exteriority, and falls back into 'the simple contiguity of things' (4). Here *Mitsein* is understood as the improper face of the Anyone (*das Man*). This interpretation denotes the '*everyday* Being-with' (4) and lacks historicity. The third interpretation denotes absolute interiority or 'a single communal *Dasein* beyond the singulars' (4). According to Heidegger, this interpretation presents the proper face of the people (*das Geschehen der Gemeinschaft, des Volkes*), and moreover denotes historicity and a 'union through destiny' (1). The second interpretation, which is neither premised on the exteriority nor the interiority of the *with*, unfortunately remains underdeveloped in Heidegger's work, with the consequence that the seemingly essential *with* is effaced. In order to see why this is the case, it is useful to follow Nancy in his analysis of Heidegger's view of death and community.

Heidegger identifies two modes of *Dasein's* being-towards-death: the improper mode and the proper mode. As regards the former, the Anyone dies an improper death for the reason that death remains an exterior experience, in which nothing is shared. The Anyone thus dies commonly, in that—as explained by Nancy (9)—'each one remains either at the mercy of or opened to its singular fate: a unique fate insofar as it is one's own death, but a banal fate insofar as it is the common cessation of life'. In contrast to the Anyone, the death of the people is characterised by its non-everydayness, in that the people have been elevated to the level and the intensity of a destiny: being-towards-death no longer concerns a sole existent's ultimate possibility, but is that through which *history* happens. The people thus represent the proper mode of dying, since death is the '*common* of a *community*' (9). On the basis of these two interpretations of death, Nancy (10) argues that death disappears twice:

> once as a common demise which remains external to the Being-delivered-over to the ultimate possibility of existing, and again according to the sublimation that the *common destiny* operates on individual death.

Nancy further argues that on both counts the *with* is effaced. In the improper mode, 'the essentiality of the *with* is dissolved' and thus becomes 'impossible' (10). This is because beings remain absolutely exterior to one another (subjects are hermetically sealed-off from one another), and death is nothing other than 'the corpse… return[ing] to the sheer material juxtaposition of things' (8). In the proper mode, 'the essentiality of the *with* determines and potentializes itself' and thus becomes 'hyperpossible' (10). This means that death is 'sublimated, sublated, or heroicized' (11) in destiny, with the consequence that being is robbed of the experience of a shared dying. Being-with is thus sacrificed for the *We*, i.e. the common subject of history.

Nancy writes that it was against his own intentions that Heidegger managed to either erase or dialecticise the existent. In another context, Nancy attributes the reason for this to the fact that '*Dasein's* "being-towards-death" was never radically implicated in its being-with—in *Mitsein*' (Nancy 1991b: 14). Nancy further states that 'it is this implication that remains to be thought' (14) and it is exactly this task that he takes upon himself by re-opening the space for a *with* that has remained 'hidden, lost or suppressed between the *Anyone* and the *people*' (Nancy 2008: 5).

5.4.2 Being as Being-with and the Question of Community

Nancy develops a non-essentialist understanding of the *with* or what Bataille (1976: 300) calls 'the naked 'with'' at the hand of the middle interpretation of *Mitdasein*. He defines being-there-with in terms of 'the common as the sharing of properties (relations, intersections, mixtures)' (Nancy 2008: 4). Nancy elaborates (10) in writing that:

> The *Mitdasein* must determine the *with* as the proximity (contiguity and distinction) of multiple *theres*, thereby giving us the following to consider: multiplicity is not an attribute extrinsic to *Dasein*, since the concept of the *there* implies the impossibility of a unique and exclusive *there*. A *there* can only be exclusive… if it equally includes a multiplicity of other *theres*.

Being (understood in relational and dynamic terms) is thus constituted by this non-essential *with*, and our being-in-common is the consequence of the fact that I never appear alone in the world. Each appearance is always already a compearance, and each death is a shared death, 'a death between us' (13). Nancy's social ontology not only leads to a rethinking of death, but also stimulates a new thinking of community and the space of the political.

As concerns the question of community, Heidegger's sublimation of being in terms of a 'destinal unity' (13) ultimately betrays a totalitarian thinking, since the *with* is essentialised as a common *We*. Nancy argues that this desire for a common identity not only defines totalitarian regimes like Nazi Germany, but underscores the whole of Western culture, to the extent that there remains a longing for 'a lost age in which community was woven of tight, harmonious, and infragible bonds and in which above all it played back to itself, through its institutions, its rituals, and its symbols' (Nancy 1991b: 9). Nancy refers to this closed conception of community as *immanentism*. Immanentism is the schema whereby the '[c]ommunity… becomes the production and the appropriation of a pre-given identity' (Devisch 2000: 247), and in which both self and other are defined in terms of the 'Economy of the Same' (245). In resurrecting a non-essentialist *with,* Nancy attempts to 'unwork' (241) this closed conception of the community, and thereby to define community in a non-immanent and non-totalitarian way.

It is interesting to note the parallels between Levinas and Nancy on this point: both identify a totalitarian logic at work in Heidegger's oeuvre (albeit that they differ regarding the source of totalitarianism[10]), and both affirm the need to create an opening in an otherwise closed conception of being. For Levinas, this opening comes to the fore in the difference between self and Other; whereas for Nancy, the relational structure of our being-in-common creates this opening.

Being-in-common or being-with others allows for a type of transcendence in immanence (what Nancy (1993) calls *transimmanence*), in that being is related to something other than its own self. In contrast to Heidegger's views on *Dasein* (which, on the final count, point to an essentialist *Dasein*), being-with is 'essentially equipriomordial in the existent' (11). Another way of saying this is that our ontology is constituted by the *with*. Devisch (2000: 244; 245) summarises as follows:

> For Nancy, 'being is 'with'. The primal ontological conditions of our community are not conceived as the One, the Other or the We, but as the 'with', 'relationality', and the 'between'. The question of being (*Seinsfrage*) is therefore the question of being-with (*Mitseinsfrage*)… The way Nancy tries to articulate our single being in the world transforms [Descartes'] *ego sum* into an *ego sum expositus* [I am exposed] or (what is the same thing) a *nos sumus* [we are].

It is important to note that this mutuality between beings implied by Nancy's radical views on relationality, is one of 'concordant discordance' (Nancy 2001: 118; trans. Watkin). This is because 'what the singularities have in common is their incommensurability' (Watkin 2007: 57). In other words, we share a 'non-essentialised, non-localizable ontology' (57), since presence (i.e. the being-in-common with) constitutes a movement wherein self never becomes self to begin with and wherein 'togetherness is otherness' (61). This view of presence needs further elaboration, since it is helpful in clarifying Nancy's understanding of community as 'a dynamic movement of sharing instead of an enclosure of already constituted subjects collectively owning a common subject' (Devisch 2000: 249).

In the preface of *Who comes after the subject?*, Nancy (1991a) writes that the existent is that which assumes a place in both space and time. The place is defined as 'a spacing that allows that something *come* into presence' which also means that '[p]resence *takes place*' (7). Since coming into presence happens in a plural manner—i.e. it is the 'shared taking place of all places'—presence can never be 'to *itsself*' (7). Presence is therefore that which we all have '*in* common' (8), where 'being 'is' the *in*' (Watkin 2007: 53). Beings thus *compear* in a presence that never stops coming or arriving. This means that 'the coming into presence is plural, "in each case ours" as much as "mine"' which also implies that 'the plural coming is a singular coming' (8); it is 'each time an other, each time with others' (Nancy 2000: 119).

Nancy's conception of presence is similar to, although more radical than, Levinas's *hypostasis*; which, to recall, is the event by which the existent comes into presence by defining itself against the anonymous current of being (i.e. existence or

---

[10] To recall, Levinas questions Heidegger's ontico-ontological distinction, arguing that Being is nothing other than a collection of beings.

the *il y a*). Furthermore, the encounter with the Other represents another moment in Levinas's work where time is interrupted, in that this encounter frees the subject from her solipsistic universe and thus serves to *presence* the ethical experience. Indeed, Diane Davis (2010) argues that the Nancean event wherein the simultaneous exposure to the other takes place is nothing other than the Levinasian encounter with the face of the Other. Derrida, in turn, refers to the event as the moment in which *limit* or *quasi-transcendental* terms (such as justice, hospitality, and forgiveness) find their expression before being taken up within a utilitarian, exchange economy. On all three counts, the subject only truly becomes in time, and within a specific context.[11]

Yet, despite this similarity, Nancy's emphasis on a shared presence serves to replace the question of self and other (which is still prized in the work of Levinas and Derrida) with the question of community. As a result, Nancy understands being as self-as-relation rather than self-in-relation (Watkin 2007). This implies that a Nancean ethics is characterised by a mutuality and a sharing, rather than by an *aporia* (Derrida) or a paralysing (in)difference (Levinas) between self and other. Christopher Watkin further argues that Nancy's ethics of mutuality constitutes 'a potent solidarity' (61), which is neither based on an essence or shared value (i.e. a communitarian position), nor on a social contract. In both of these latter interpretations, the self is understood as related to, but also separated from, other beings. Nancy, on the other hand, understands solidarity in terms of the self as being-with others. This is true in so far as 'I am singular plural relation' and therefore share my finitude and incommensurability with other bodies.

One point of critique that can be lodged against this conception of identity is that the dissolution of the distinction between the self and the other means that not only my face, but also the face of the other is lost. This loss is palpable in the post-structural philosophical discourse (where the face of the other previously dominated). However, in the complexity discourse, the face never made an appearance. This lends support to the view that the complexity conception of identity may share a greater affinity with Nancy's project than with Derrida's, because—unlike Nancy—Derrida is still concerned with the unique and singular gaze of the other (even if such an other is a cat). In this regard, it is worthwhile to note that Nancy's conception of being-with or self-as-relation is compatible with certain elements of Cilliers'

[11] Although the moment of coming into presence is not as highly valued in the complexity discourse, identity is nevertheless viewed as an emergent property to the extent that 'it develops and transforms as a result of the play of differences which constitute it' (Cilliers 2010a: 7). Due to dynamic feedback loops and non-linear interactions, emergence should also not be viewed as a progressive or incremental process. Nancy's definition of the we 'as 'each one' (chaque un) and 'each time' (chaque fois)' (Watkin 2007: 55), as well as Derrida's concepts of *différance* and *iterability* (where the latter implies a view of identity that ties repetition or sameness to difference or alterity) also guard against this teleological view. Becoming in time thus renders the subject as fundamentally irreducible to itself. That the subject's sense of identity is contingent on context also reaffirms the view that identity cannot be a pre-given or complete construct. Rather, we are the product of our histories, our relationships, and our current contexts. As such, we are forced to always renegotiate our identities in practice, and with one another.

view of identity, specifically as concerns Cilliers' emphasis on the *relations* between components as the primary unit of analysis in complex systems, his view that richly-connected systems may consist only of boundaries, and his focus on the emergence of a complex systemic order as such. In Nancean terms, these three points translate as a concern with the *with* of being-with or the way in which beings are *in* common, the re-description of *ego sum* as *ego sum expositus* (which means that the identities of beings reside in their exposure to one another), and a focus on the community (the system) as opposed to the individual (the component).

Regardless of which position one endorses, it is clear that—given the above accounts of subjectivity and identity—ethics (and the related constructs of moral agency, responsibility, and accountability) should not be premised on a view of the subject that had been thoroughly debunked during the course of the twentieth and twenty-first centuries. To think of our responsibilities in terms of the identity predicates of an isolated and anthropocentric Cartesian subject, leads to the belief that we are able to change both our individual fortunes and the fortune of the world by relying solely on our own agency and mental faculties. At this point in the argument, it should be clear that this cannot be the case. The question that thus aries is how to think of our ethical duties and responsibilities in a post-Cartesian world, and it is to this issue that we turn next. The analysis proceeds at the hand of the example of the foreigner, stranger, or intruder.

5.5 The Foreigner That Intrudes

In exploring our duties to the foreigner, we begin with Levinas's view of the subject. As previously argued, for Levinas, the solipsistic ego is decentred in the encounter with the face of the Other. This encounter is characterised by an asymmetrical relationality, which refigures the subject's identity in terms of the imperative to respond to the Other. In a course that Levinas delivered in 1975–1976, he summarises this view as follows:

> Someone who expresses himself in his nakedness—the face—is in fact one to the extent that he calls upon me, to the extent that he places himself under my responsibility: I must already answer for him, be responsible for him. Every gesture of the Other was a sign addressed to me... The Other who expresses himself is entrusted to me (and there is no debt with regard to the Other—for what is due cannot be paid; one will never be even). The Other individuates me in my responsibility for him (Levinas 1993: 25–26; trans. Brault and Naas).

For Levinas, the Other is the stranger, and my mode of being is that of being-for-the-Other. Peperzak (1993) translates this as meaning that my home, my labour, and the fruits of my labour only receive their true meaning when put into the service of, and offered to, the Other for whom I am infinitely responsibly. In Perperzak's (24–25) words: 'I must feed my body and arrange my house in order to receive the foreigner knocking at my door; if I possess a home, it is not for me alone.' Derrida's understanding of the host as hostage (which was introduced in Sect. 4.5.2, and which will again be returned to in this discussion) has its origins in Levinas's work. In his view, the claims of the Other necessarily precede the subject's own claims, in that the subject is first-and-foremost at the Other's disposal. Levinas's great friend, Maurice Blanchot (1993: 51–52), remarks on the radicality of this account, in stating that:

> In Emmanuel Levinas's book [*Totality and Infinity*]... we are called upon to become responsible for what philosophy essentially is, by welcoming, in all the radiance and infinite exigency proper to it, the idea of the Other, that is to say, the relation with *autrui.* It is as though there were here a new departure in philosophy and a leap that it, and we ourselves, were urged to accomplish.

Derrida takes this leap that Levinas facilitates in so far as he raises the problematic of the foreigner (a problematic which is alternatively expressed in terms of the question of the animal). Like Levinas, Derrida views the question of the foreigner (or the other) as intimately linked to the subject's identity. In the opening page of *Of Hospitality*, he analyses this question as follows:

> There is... a question of the foreigner... But before being a question to be dealt with, before designating a concept, a theme, a problem, a program, the question of the foreigner is a question *of* the foreigner, addressed *to* the foreigner... As though the foreigner were being-in-question, the very question of being-in-question, the question-being or being-in-question of the question. But also the one who, putting the first question, puts me in question (Derrida 2000: 3).

The question of the foreigner is thus inseparable from the question of being, and moreover from the question of *my* being. To understand how Derrida develops this idea, it is useful to revisit the *aporetic* logic that informs his view of identity.

On the one hand, our identities are relatively stable and well-demarcated from one another. Our identities emerge in differential networks, whereby all identities are both related to one another, but also distinguished from one another. Cilliers (2010b: 61) calls this the law of meaning, which he formulates as follows: '*without constrained difference and repeatable identity, there can be no meaning'*. Since identity is the product of difference, we are forced to contend with these differences, rather than attempt to assimilate them. Yet, since identities are related to one another, they also influence one another. In this regard, it is useful to recall Derrida's discussion on the foreigner or the *Xenos*, in which he argues that the foreigner is allowed into the city, but is never fully assimilated by the apparatus of the city. The identity accorded to the foreigner is determined by the laws of hospitality, which also serve to distinguish the foreigner from the citizen (and, in so doing, grounds the identity of the citizen). Mark Westmoreland (2008: 2) explains as follows:

> The foreigner occupied an integral space within the city. Indeed, the foreigner was essential because he provided that to which citizens could compare themselves. From a phenomenological standpoint, one could claim that one's identity is only understood in relation to others. Citizens understand themselves in relation to others, to foreigners. "We are not those sorts of people. We are citizens."

On the other hand, identity is undecidable; and the identity of the other is characterised by an absolute alterity that can neither be conceptualised nor assimilated (even in part), and that consequently begs the question, 'Who is the other?'. This absolute alterity not only governs over identity, but also over ethics, responsibility, and justice, in that it guards against the systematisation of these non-concepts. On this reading, identity can only be grappled with by means of unconditional hospitality, which is both 'inconceivable and incomprehendable' (Derrida 2002d: 362). Derrida (1999) equates this unconditional hospitality with ethics[12] (understood as a radical ethics of alterity). In absolute or unconditional hospitality, I am denied any clue as to the stranger's identity, including his name. In Westmoreland's (2008: 5) words: 'absolute hospitality relies upon the deafening silence between the *ipse* and the other.' And yet, the aneconomic force of the stranger, which serves to deconstruct the home, does have an affect on my identity and may even destroy my identity.

Levinas (1993: 26) maintains that the Other transforms my subjecthood in 'individuat[ing] me in my responsibility for him'. Similarly, Derrida argues that, in welcoming the stranger, the self is interrupted. Instead of receiving the stranger, I am received by him. In deconstructing my home, the stranger challenges the conceptual order that determines my identity. The stranger thus denies me mastery of my home, and by implication, my life. Derrida (2000: 125) argues that 'the master of the house is at home, but nonetheless he comes to enter his home through the guest—who comes from outside.' As such, the host becomes the guest or hostage of the other. Or, as stated by Westmoreland (2008: 7), '[t]he *ipse* as host has been interrupted.'

Like Derrida and Levinas, the notions of otherness and selfhood have greatly interested Nancy. He has explored these issues at length, particularly in terms of the implications that they hold for our understanding of community and multiculturalism. Although Nancy does not refer specifically to the relation between otherness and identity as an *aporia*, he does seem to be sensitive to the issues that this *aporia* raises. In particular, he also grapples with the difficulty of respecting the foreigner's differences, whilst at the same time relating to the foreigner as someone who shares my world. One example of Nancy's engagement with this issue is a short film called 'Vers Nancy', which forms part of a collection of short films, titled *Ten Minutes Older: The Cello* (2002). 'Vers Nancy' is directed by Claire Denis. She is an avid follower of Nancy, and—like Nancy—explores 'the deeply perplexing questions of identity and alienation, assimilation and rejection, desire and fear' (Beugnet 2008: 31). This black-and-white short depicts Nancy having a conversation with Ana Samardzija (a student of his) on a train journey. Martine Beugnet (33) writes that during their conversation:

---

[12] As previously stated, unconditional hospitality also raises questions regarding the anthropological dimension of hospitality, specifically as regards the problem of who is accorded the right to hospitality. In this regard, Derrida (2010: 4) asks:

> what can be said of, indeed can one speak of, hospitality towards the non-human, the divine, for example, or the animal or vegetable; does one owe hospitality, and is that the right word when it is a question of welcoming—or being made welcome by—the other or the stranger [*l'étranger*] as god, animal or plant, to use those conventional categories?

> they evoke that inherent but necessary contradiction that lies at the heart of the construction of individual and collective identities: the existence of an 'other' as the very fundament for self-identity, and the need to define oneself through difference, with and against the other, simultaneously denying one's own internal fragmentation.

As concerns the above contradiction, Nancy argues that it is impossible to fully assimilate the foreigner without denying his existence. The true skill is to learn how to be with the foreigner. Otherwise put, the experience of foreignness is 'the difficult experience of being with the intruder, of being intruded upon' (33).

'Vers Nancy' was followed by another piece inspired by the work of Nancy, titled *L'Intrus* in which Denis (2004) adopted Nancy's autobiographical text by the same name, and translated the feel of this text into a film narrative. *L'Intrus*, which deals with Nancy's experience of having a heart transplant, is a remarkable text for the reasons that it provides 'a thought-provoking analogy between the physical and psychological implications of the transplant and the fear of being intruded upon' (34). It also blends together a very personal account with a philosophical reflection on the nature of foreignness, and the challenges it poses for current geo-politics. The balance between the personal and the theoretical is further facilitated by switching between a first-person account and the passive voice (Beugnet 2008).

Nancy's text opens with the problematic of the intruder. He writes that:

> The Intruder (*l'Intrus*) enters by force, through surprise or ruse, in any case without the right and without first having been admitted. There must be something of the *intrus* in the stranger; otherwise, the stranger would lose its strangeness: if he already has the right to enter and remain, if he is awaited and received without any part of him being unexpected or unwelcome, he is no longer the *intrus,* nor is he any longer the stranger. It is thus neither logically acceptable, nor ethically admissible, to exclude all intrusion in the coming of the stranger, the foreign (Nancy 2002: 1).

As with Derrida, Nancy thus identifies the necessity of a certain undecidability when dealing with the stranger. The stranger must be unexpected, and because of the threat he holds, unwelcome. And, yet, it is exactly on the grounds that he is unwelcome that an ethics or hospitality is possible. Nancy juxtaposes this ethics to the stranger (Derrida's unconditional hospitality) with 'moral correctness [*correction morale*]' (2) (Derrida's conditional law of hospitality), which 'assumes that one receives the stranger by effacing his strangeness at the threshold' (2). Nancy insists that one cannot receive the stranger without experiencing his intrusion.

After the introduction to the central theme of the text, Nancy describes his shock at learning that his heart was defective and needed to be replaced. In gaining this knowledge, Nancy writes that his heart, which—until that point—had been 'as absent to [his reflection]… as the soles of [his] feet walking' (3), became a stranger to him. His defective heart was threatening or intruding upon his body. It therefore had to be extruded. Nancy (4) describes his experience as follows:

> My heart was becoming my own foreigner—a stranger precisely because it was inside. Yet this strangeness could only come from outside for having first emerged inside. A void suddenly opened in my chest or my soul—it's the same thing—when it said to me: "You must have a heart transplant…"

Note how similar this formulation is to Derrida's (2000: 125) description of the stranger who upsets the identity of the host: 'the master of the house is at home, but nonetheless he comes to enter his home through the guest—who comes from the outside.' What both Nancy and Derrida are referring to is that an intrusion can only be felt if it upsets and destabilises the very core of our being (my home, my chest, my soul).

Next, Nancy describes the difficult process of being intruded upon by the medical and administrative machinery that make a heart transplant possible. In this regard, he writes: 'It's not hard to imagine the complexity of the strange ensemble that in this way intervenes in what is most intensely "me."' (Nancy 2002: 4); and: 'From the first, my survival is inscribed in a complex process woven through with strangers and strangeness' (5). This experience leads Nancy to reflect on a number of issues, including: the right measures and justness in according organs to candidates (given that the demand outstrips the supply); the desire for immortality and the way in which our technological developments prolong life, but also increasingly isolate us from death and from nature; the definition of life 'proper' (which is contingent on our essential organs, but certainly does not reside in them); organ donation and the sharing of life/death; and, the strangeness that manifests due to the possibility of having one's body reject a foreign organ.

With regard to this latter point, Nancy (9) writes that:

> The possibility of rejection establishes a strangeness that is two-fold: on the one hand, the foreignness of the grafted heart, which the host body identifies and attacks insomuch as it is foreign; and, on the other, the foreignness of the state that the medical regimen produces in the host body, to protect the graft against rejection.

In terms of the medical regimen, treatments are given to a transplant patient in order to lower his immunity or his immunity's identity (what Nancy calls his *physiological signature*). This is done so that his immunity (which is becoming depleted and hence foreign to itself) may accept the foreign organ. Hence strangeness proliferates: 'As soon as intrusion occurs, it multiplies, making itself known through its continually renewed internal differences' (9). Nancy equates immunity with identity, arguing that '[t]o reduce the one is to reduce the other', to the point that the experience of strangeness becomes ordinary. 'This' he writes, 'is expressed through a constant self-exteriorization: I must be monitored, tested, measured' (9) in order to expose those enemies and viruses (i.e. those intruders) lurking within me. As concerns his own identity, Nancy writes:

> "I" has clearly become the formal index of an unverifiable and impalpable system of linkages. Between my self and me there has always been a gap of space-time: but now there is the opening of an incision ['I am closed open' (10)] and an immune system that is at odds with itself, forever at cross purposes, irreconcilable (10).

Nancy continues to describe how his feelings of self-alienation are further heightened by the onset of cancer (the result of his lowered immunity), and a second regimen of invasive treatments. Here, he attributes the ever-growing distance from himself to the experiences of pain and suffering. 'Very quickly,' he writes, 'one is no more than a slackening, floating strangeness, suspended between poorly identified states, between sufferings, incapacities, lapses' (11). As such, '"I" end/ ends up being no more than a tenuous thread—from pain to pain, strangeness to strangeness' (12).

In the last page of the text, Nancy reflects in more detail on the implication that his experiences hold for his identity, but also for identity as such. The intruder (in his case, his illnesses and the medical interventions that followed) reveals the truth of the subject, and this truth lies in the subject's 'exteriority and excessivity: its infinite exposition' (13). Who are we? We are the intruder (in Nancy's case, the wire, the stitches, the screws, the plates, and the medications that literally hold him together). The intruder prevents us from clearly identifying or demarcating the subject. On this point, parallels can again be found with Derrida (1995: 285), who writes that 'we never know, and never have never known, how to cut up a subject'. Nancy (2002: 13) writes that our very ontology is defined by intrusion by the other:

> The *intrus* is no other than me, my self; none other than man himself. No other than the one, the same, always identical to itself and yet that is never done with altering itself. At the same time sharp and spent, stripped bare and over-equipped, intruding upon the world and upon itself: a disquieting upsurge of the strange, *conatus* of an infinite excrescence.

The fact that the other or the intruder is *a priori* identified with the subject, or as part of the subject, means that—unlike in the philosophy of Levinas—I am not the Other's guardian. Ethics does not amount to protecting the Other but to learning to live together with the other in an infinite excrescence.

5.6 Conclusion

The post-structural positions explored in this chapter lead to very different views of responsibility than the view that comes to the fore in the Cartesian model. In post-structuralism, responsibility is not something that is intentionally and rationally assumed once we have made the necessary calculations. Rather responsibility is the outcome of our *mode* of being in the world, which is characterised by an exposure to otherness. Since responsibility is demanded by our phenomenological condition, assuming responsibility is also no longer a conscious choice. For Levinas, the face of the Other compels us to accept responsibility; whereas for Derrida, our being together in the world generates an infinite responsibility for our actions (even though these actions are neither guided by a transparent rationality nor lead to causal outcomes that can be anticipated in advance). Derrida's understanding of identity (which problematises the distinction between self and other) straddles the divide between Levinas and Nancy, in so far as one can identify an overlap with Levinas's ethics of alterity and Nancy's view of self-as-other. Both Derrida and Nancy support the view that the stranger resides in the very heart of being (in Nancy's case, the stranger is his heart) and learning to live with one's self is equivalent to learning to live with the other, who dwells in one's heart, and who resists definition and appropriation. Ethics (i.e. the difficult task of reconciling oneself to that which/who is unwelcome and unexpected) therefore cannot be severed from being. However, for Nancy the notion of self-as-other is the basis for mutuality (in that the experience of otherness is exactly what we have in common), whereas for Derrida otherness (understood here in terms of alterity, as opposed to difference) is the result of the aneconomic force. Rather than serve as the basis for a conception of being as being-with, the aneconomic force threatens to ruin the economic dimension that allows us to define, and identify with, being as such.

The shift from self-in-relation (Levinas) to self-as-relation (Nancy) leads to interesting ethical implications. Watkin (2007) contends that Nancy's singular-plural (or the non-essentialising mode of being introduced by the *Mitdasein*) allows us to maintain an ethical position that does not immediately translate into the imperative of infinite responsibility for the Other (or—one can add—to the annulment of otherness through action, as is the case in Derrida's conception). In contrasting Nancy's *ethics of mutuality* to Levinas's position, Ian James (2005) remarks that '[t]he ethical relation is not "passed over" in Nancy, it is simply thought of differently as a relation of being side-by-side rather than as "otherwise than being" of transcendence in the face-to-face' (343). Nancy's ethics thus entails a grappling with being-as-otherwise, rather than a continual engagement with that which is otherwise than being.

This difference has very real practical consequences, as demonstrated by Watkin (2007). He argues that 'the fire of conflict is stoked by the inability to disengage from the dichotomy of self-identity and absolute alterity', upon which both 'the infinite generality of globalisation' and 'the bellicose intensity of a fundamentalist essentiality' are based (Watkin, 52). Whereas the former represents a hegemonic sameness, the latter is defined by a dichotomy between self and other. Although Levinas's position guards against these outcomes by stressing both that the relation with the other is first-and-foremost an *ethical* relation, it is possible to see how the general schema of self/other can be abused and hence lead to a negative politics of identity. Nancy's radicalisation of this problematic, which centres on the difficult question of sameness given 'the non-coincidence of the Same with itself' (Zizek 2006: 36), does not allow for a straightforward polarisation between self and other.

Despite the differences between these positions, all three positions present us with examples of modes of being in the world. The conception of ethics as a mode of being supersedes traditional normative categories (in which right and wrong actions are determined in advance), and instead embodies a type of *ethics of living*. The latter is something that cannot be theoretically taught and does not lend itself to categorical rules. Rather, it encapsulates a certain attitude towards life, and therefore constitutes a *praxis* of living. The nature of this praxis will be explored in more detail in the following chapter.

References

Adams, C.J. 1990. *The sexual politics of meat*. New York: Continuum.

Badiou, A. 2001. *An essay on the understanding of evil*. Trans. P. Hallward. London: Verso.

Bataille, G. 1976. *Oeuvres complètes VIII*. Paris: Gallimard.

Bernasconi, R. 1987. Deconstruction and the possibility of ethics. In *Deconstruction and philosophy: The texts of Jacques Derrida*, ed. J. Sallis, 122–139. Chicago: University of Chicago Press.

Beugnet, M. 2008. The practice of strangeness: *L'Intrus*—Claire Denis (2004) and Jean-Luc Nancy (2000). *Film-Philosophy* 12(1): 31–48.

Blanchot, M. 1993. *The infinite conversation*. Trans. S. Hanson. Minneapolis: University of Minnesota Press.

Bordo, S. 1987. *The flight to objectivity: Essays on Cartesianism and culture*. New York: State University of New York Press.

Calarco, M. 2004. Deconstruction is not vegetarianism: Humanism, subjectivity and animal ethics. *Continental Philosophy Review* 37(2): 175–201.

Campbell, E. 2003. Our world: An interview with J.-L. Nancy, with translator's introduction, trans. E. Campbell. *Angelaki: Journal of the Theoretical Humanities* 8(2): 43–54.

Cilliers, P. 1998. *Complexity & postmodernism*. London: Routledge.

———. 2010a. Difference, identity and complexity. In *Complexity, difference and identity*, ed. P. Cilliers and R. Preiser, 3–18. Dordrecht: Springer.

---------. 2010b. Difference, identity, and complexity. *Philosophy Today* Spring 2010: 55–65.

Cilliers, P., and T. de Villiers. 2000. The complex I. In *The political subject*, ed. W. Wheeler, 226–245. London: Lawrence & Wishart.

Cilliers, P., T. de Villiers, and V. Roodt. 2002. The formation of the self. Nietzsche and complexity. *South African Journal of Philosophy* 21(1): 1–18.

Clark, D.L. 2004. On being "the last Kantian in Nazi Germany": Dwelling with animals after Levinas. In *Postmodernism and the ethical subjects*, ed. B. Gabriel and S. Ilcan, 41–74. Kingston and Montreal: Queen's UP.

Cornell, D. 1992. *The philosophy of the limit*. New York: Routledge.

Critchley, S. 1999. *The ethics of deconstruction: Derrida and Levinas*. Edinburgh: Edinburgh University Press.

———. 1999b. *Ethics, politics, subjectivity. Essays on Derrida, Levinas and contemporary French thought*. London: Verso.

Culler, J. 1983. *On Deconstruction: Theory and criticism after structuralism*. London: Routledge.

Davis, D. 2010. *Inessential solidarity: rhetoric and foreigner relations*. Pittsburgh: University of Pittsburgh Press.

Denis, C. 2004. *L'Intrus*. France.

Derrida, J. 1976. *Of grammatology*. Trans. G. Spivak. Baltimore: Johns Hopkins University Press.

———. 1978. Violence and metaphysics: An essay on the thought of Emmanuel Levinas. In *Writing and difference*. Trans. A. Bass, 79–102. Chicago: University of Chicago Press.

———. 1987a. *Of spirit: Heidegger and the question*. Trans. G. Bennington and R. Bowlby. Chicago: The University of Chicago Press.

———. 1987b. On reading Heidegger: An outline of remarks on the Essex colloquium. *Research in Phenomenology* 17: 171–185.

--------. 1982. Différance. In *Margins of philosophy*. Trans. A. Bass, 1-28. Chicago: University of Chicago Press.

———. 1988. Afterword. In *Limited Inc.*, ed. G. Graff, trans. S. Weber, 111–160. Evanston: Northern Western University Press.

———. 1995. '"Eating well," or the calculation of the subject', interviewer J.-L. Nancy, trans. P. Connor and A. Ronell. In *Points...Interviews, 1974–1994*, ed. E. Weber, 255–287. Stanford: Stanford University Press.

———. 1999. *Adieu*. Trans. P.-A. Brault and M. Naas. Stanford: Stanford University Press.

———. 2000. *Of hospitality.* Trans. R. Bowlby. Stanford: Stanford University Press.
———. 2002a. The Animal that therefore I am (More to follow). Trans. D. Wills. *Critical Inquiry* 28(2): 369–418.
———. 2002b. Ethics and politics today. In *Negotiations: Interventions and interviews 1971–2001,* ed. and trans. E. Rottenberg, 295–314. Stanford: Stanford University Press.
———. 2002c. Force of law: The "Mystical Foundations of Authority". In *Acts of religion,* ed. and trans. G. Anidjar, 228–298. New York: Routledge.
———. 2002d. Hospitality. In *Acts of religion*, ed. and trans. G. Anidjar, 356–420. New York: Routledge.
———. 2003. And say the animal responded?' In *Zoontologies: The question of the animal,* ed. C. Wolfe, trans. C. Wolfe and D. Willis, 121–146. Minnesota: University of Minnesota Press.
———. 2005. The principle of hospitality. *Parrallax* 11(1): 6–9.
———. 2009. *The animal that therefore I am.* Trans. D. Wills. New York: Fordham University Press.
———. 2010. Hostipitality. *Angelaki: Journal of the Theoretical Humanities* 5(3): 3–18.
Derrida, J., and E. Roudinesco. 2004. Violence against animals. In *For what tomorrow...A dialogue*. Trans. J. Fort, 62–76. Stanford: Stanford University Press.
Descartes, R. 1960. *Discourse on method and other writings*. Trans. A. Wollaston. Middlesex: Penguin Books Ltd.
Devisch, I. 2000. A trembling voice in the desert: Jean-Luc Nancy's rethinking of the space of the political. *Cultural Values* 4(2): 239–256.
Gasché, R. 1994. *Inventions of Difference.* Cambridge, MA: Harvard University Press.
Glendinning, S. 1998. *On being with others: Heidegger-Derrida-Wittgenstein*. London: Routledge.
Heidegger, M. 1962. *Being and time.* Trans. J. Macquarrie and E. Robinson. New York: Harper and Row.
Howells, C. 1998. *Derrida: Deconstruction from phenomenology to ethics*. Cambridge: Polity Press.
James, I. 2005. On interrupted myth. *Journal for Cultural Research* 9(4): 331–349.
Krell, D.F. 1988. Engorged philosophy: A note on Freud, Derrida, and *Différance*. In *Derrida and Différance*, ed. D. Wood and R. Bernasconi, 6–11. Evanston: Northwestern University Press.
Levinas, E. 1969. *Totality and infinity.* Trans. A. Lingis. The Hague: Martinus Nijhoff.
———. 1988. *Existence to existents.* Trans. A. Lingis. Dordrecht: Kluwer Academic Publishers.
———. 1993. La Mort et le temps. In *Dieu, la mort et le temps.* Paris: Grasset.
———. 1998. *Otherwise than being, or beyond essence.* Trans. A. Lingis. Pennsylvania: Duquesne University Press.
Levinas, E., and R. Kearney. 1986. Dialogue with Emmanuel Levinas. In *Face to face with Levinas*, ed. R.A. Cohen, 13–34. New York: State University of New York.
Lingis, A. 1988. Translator's introduction. In *Existence to existents*, ed. E. Levinas, 7–14. Dordrecht: Kluwer Academic Publishers.
Morin, E. 1980. *La Méthode, v. 2: La Vie de la vie*. Paris: Seuil.
Nancy, J.-L. 1991. Introduction. In *Who comes after the subject?*, ed. E. Cadava, P. Connor, and J.-L. Nancy, 1–8. New York: Routledge.
———. 1991b. *The inoperative community*, ed. P. Connor, trans. P. Connor, L. Garbus, M. Holland and S. Sawhney. Minneapolis: Minnesota University Press.
———. 1993. *Le sens du monde*. Paris: Galiléé.
———. 2000. *Being singular plural.* Trans. R.D. Richardson and A.E. O'Byrne. Stanford, CA: Stanford University Press.
———. 2001. *La pensée dérobée.* Paris: Galilée.
———. 2002. *L'Intrus.* Trans. S. Hanson. Michigan: Michigan State University Press.
———. 2008. The being-with of being-there. Trans. M.E. Morin. *Continental Philosophical Review* 41: 1–15.
———. 1989. *Beyond good and evil: Prelude to a philosophy of the future*. Trans. W. Kaufmann. New York: Vintage Books.

Peperzak, A. 1993. *To the other: An introduction to the philosophy of Emmanuel Levinas*. Indiana: Purdue University Press.
Saussure, F. 1960. *Course in general linguistics*, ed. C. Bally and A. Sechehaye, trans. W. Baskin. London: Peter Owen.
Seabright, M.A., and L.B. Kurke. 1997. Organisational ontology and the moral status of the corporation. *Business Ethics Quarterly* 7(4): 91–108.
Spivak, G.C. 1994. Responsibility. *Boundary 2* 21(3): 19–64.
Various Directors. 2002. *Ten minutes older: The cello*. UK/Germany/France.
Watkin, C. 2007. A different alterity: Jean-Luc Nancy's 'singular plural'. *Paragraph* 30(2): 50–64.
Westmoreland, M.W. 2008. Interruptions: Derrida and hospitality. *Kritike* 2(1): 1–10.
Williams, C. 2001. *Contemporary French philosophy: Modernity and the persistence of the subject*. London: Athlone Press.
Wolff, E. 2010. Rethinking the conditions for inter-cultural interaction: A commentary on Levinas' humanism of the other. *Taiwan Journal for South East Asian Studies* 7(2): 113–147.
Wood, D. 1999. *Comment ne pas manger*—Deconstruction and humanism. In *Animal others: On ethics, ontology and animal life*, ed. H.P. Steeves, 15–35. Albany: State University of New York Press.
Zizek, S. 2006. *The parallax view*. Cambridge, MA: MIT Press.

Chapter 6
The Ethics of Living

It is clear, however, that ethics has nothing to do with punishment and reward in the usual sense of the terms. Nevertheless, there must indeed be some kind of ethical reward and punishment, but they must reside in the action itself.

- Ludwig Wittgenstein (1921), Logisch-Philosophische Abhandlung

Abstract
In this chapter, it is argued that both the philosophical and complexity positions that are drawn on in this work necessitate an ethics of living, understood in terms of ethical praxis.

In the traditional understanding of moral theory, the theorist seeks to defend a priori moral obligations (deontology) or formulate a moral purpose (teleology), both of which can be translated into specific principles and rules. If however one accepts complexity (including moral complexity), then it stands to reason that meaning (including moral meaning) cannot be fixed in advance. As a result, morality (which is traditionally understood in terms of a normative system dictating right action) is reinscribed as praxis, wherein decisions and actions are not measured against pre-defined norms, but come to constitute the very subjects that undertake them.

The exact nature and significance of ethical praxis however varies according to the position studied. In this chapter, two views on praxis are presented, namely Derrida's experiential praxis and Nancy's praxis defined as an engagement in, and with, originary ethics. Whereas Derrida's praxis gives rise to an ethics of alterity, Nancy's praxis translates as the duty to make sense or the duty to respect existence. The practical implications that their insights hold for understanding, and practising, the ethics of living are also explored.

M. Woermann, *Bridging Complexity and Post-Structuralism*,
DOI 10.1007/978-3-319-39047-5_6

6.1 Introduction

Although the question of ethics was not dealt with explicitly in the previous chapters, the discussions in these chapters were nevertheless permeated with normative insights and implications stemming from both the philosophical and the complexity literature. These normative insights largely serve to dispel the fundamental assumptions upon which ethical theory is based, namely that morality is grounded in a pre-formulated purpose or that morality is grounded in duty. The former branch encompasses teleological theories, whereas the latter concerns deontological theories. The respective duty of the moral philosopher under these two conceptions of morality is either to explain how the moral purpose should be realised, or to explain how *a priori* obligation is implemented in concrete moral imperatives (van Tongeren 2008).

The most common examples of individual and institutional teleological theories are Aristotle's eudemonistic ethics and Mill's utilitarian ethics respectively. The former attempts to provide an answer to the question 'How should I live my life?'; whereas the latter is concerned with how matters should be arranged in general. In both cases however, the answer to these questions accords with a pre-formulated purpose. Kant and Rawls also attempt to provide answers to these questions, but their answers are based on an assumed fundamental obligation, which transcends contextual contingencies. As such, Kant's duty-based ethics and Rawls's contractualist ethics constitute the primary examples of individual and institutional deontological theories (van Tongeren 2008).

Positions that are sensitive to complexity and post-structural insights (as presented in the preceding chapters) reject both teleological and deontological theories, because—to repeat—it is believed that moral concerns can neither be grounded in a pre-formulated purpose nor in *a priori* obligations. The reason for these beliefs can be found in the conception of economy forwarded in Chap. 3. To recall: the regulated economy of meaning necessarily includes an aneconomic dimension, which serves to disrupt its regulation, and which defines meaning as open. Meaning, including moral meaning, therefore cannot be fixed in advance in either purpose or obligation, as doing so denies the aneconomic dimension that introduces complexity. To fix moral meaning is therefore to deny complexity, which denotes a position that holds serious ethical consequences.

There are different ways in which to conceptualise the openness of meaning. Derrida does so primarily through means of his *aporetic* logic that defines the workings of *différance*. He illustrates the undecidable nature of meaning by drawing on examples such as the gift, hospitality, and the animal; as well as by interrogating the relation between justice and law, ethics and politics, reflection and intervention, and theory and practice.[1] Nancy, on the other hand, locates the resources for non-totalising meaning in his differential view of the self and his view of transimmanence (which stem from an understanding of being as being-in-common with

[1] See Sects. 3.4.2 and 4.3.1.

others).[2] Nancy's assertions that the self can never correspond with itself (the intruder is no one other than me), and that community can never be reduced to the common *we* or destinal unity of the people, guard against a closed or economic view of meaning.

The philosophical complexity literature also supports the position that the full meaning of complex phenomena can never be grasped. This is partly because of our bounded rationality (which compels us to engage with the status and limits of our knowledge claims); but, more importantly, it also has to do with the nature of complex phenomena as such. Complex phenomena do not emerge because of a central organising principle. To paraphrase Derrida (1978), the centre does not hold. This means that even with infinite computing power, it would be impossible to uncover the base laws or grounds of complex phenomena, for the simple reason that such fundamental laws do not exist. This is not to say that lawful behaviour or localised patterns also do not exist. The point is rather that there is no *Grand Design* underlying complex phenomena. Engaging with complex phenomena therefore necessitates an 'inquiry into the relation… between the organization of knowledge and the knowledge of organization' (Morin 2008: 109).

Morin however argues that the challenge posed by complexity thinking moves beyond this general epistemological level, to influence the whole system of thought that defines our specific thinking on matters related to our practices, politics, and ethics. If we cannot fully know complex phenomena, then it means that we cannot calculate their behaviour in any deterministic fashion (in moral terms, we cannot blindly follow imperatives designed to meet a predetermined purpose or obligation). Instead, we must interpret and evaluate, and our decisions necessarily require that we grapple with the aneconomic dimension of meaning. In this regard, Derrida (1988: 116) writes that the undecidable 'opens the field of decision or of decidability… There can be no moral or political responsibility without this passage by way of the undecidable.' In a published plenary presentation, titled 'Ethics and second-order cybernetics', von Foerster (1990) makes a similar point in discussing the nature of undecidable questions. He argues that '*[o]nly those questions that are in principle undecidable, we can decide*.' The reason for this is that:

> decidable questions are already decided by the choice of the framework in which they are asked, and by the choice of rules of how to connect what we call "the question" with what we take for an "answer." In some cases it may go fast, in others it may take a long, long time, but ultimately we will arrive, after a sequence of compelling logical steps, at an irrefutable answer: a definite Yes, or a definite No.

Knowledge of complex phenomena is undecidable for the reason that there is no *a priori* framework for determining the rules that will lead to the correct answer. Our theoretical frameworks therefore cannot be justified objectively, but are, in part, based on normative judgements. This means that the ethics implied by complexity thinking is not an add-on, but inherent to any real engagement with complex phenomena. Otherwise stated, the *ethics of complexity* is a structural condition for

[2] See Sects. 3.5.3 and 5.4.2.

complexity thinking (Woermann and Cilliers 2012). In practice, this means that we should assume a critical attitude when modelling phenomena. The critical attitude amounts to both the recognition of, and engagement with, the limits of knowledge (Preiser and Cilliers 2010).

Despite the fact that complexity thinking necessarily involves ethics, engaging with complexity does not provide us with any information regarding the content of such an ethics. This is because our sense of right and wrong, good and bad, and what deserves respect and what does not, cannot be justified on teleological or deontological grounds. Moreover, a complexity-based position implies that any substantive notion of ethics must itself be subjected to a deconstruction of sorts, since our ethical models are limited and, hence, exclusionary (Cilliers 2005; Derrida 2005, 2002a, b, 1999). As such the logic that informs the *ethics of complexity* commits us to accepting the *complexity of ethics*, which is a position wherein no ethical model can be viewed as categorically binding (Woermann and Cilliers 2012).

The philosophical and complexity perspectives presented in this work prevent one from giving a substantive account of ethics. However, this does not mean that they are mute on issues of responsibility, dignity, and care for others. To the contrary, I argue that the positions forwarded in this work compel us to accept an ethics of living, which should be understood in terms of ethical praxis. The word *praxis* is derived from Aristotle, and is contrasted with poiesis. Praxis means action or doing something; whereas poiesis means making, producing, or creating something. This chapter is devoted to exploring ethical praxis, at the hand of the insights derived from Derrida's ethics of alterity and Nancy's reading of a Heideggerian originary ethics. These positions will be compared with each other, and the implications that they hold for our understanding of the role of philosophy and thought will be investigated.

6.2 Derrida: (Im)Possible Experience as Praxis

Derrida's *aporetic* logic defines moral action in terms of undecidability. In Derrida's (1999: 66) words: 'Ethics and politics… start with undecidability'. Since undecidability characterises the decision-making process, it stands to reason that the decision must be 'heterogeneous to knowledge' or must 'go beyond knowledge' (66). In terms of moral knowledge, this means that 'the distinction between good and evil doesn't depend on knowledge… we should not know, in terms of knowledge, what… the distinction between good and evil [is]' (66). To distinguish between good and evil and to assume responsibility for a decision is, Derrida confesses, 'a terrible and tragic situation in which to find oneself' (66). And yet this experience is unavoidable if one wishes to define responsible action as something other than 'a serene application of a programme of knowledge' (67).

The undecidable nature of the decision-making process—indeed, of thought itself—is, in part, due to the fact that the world is characterised by too much determinacy. In other words, when it comes to issues of justice, Derrida argues that there is always competition between two or more determined possibilities, outcomes, or duties—all of which can be equally justifiable. However, at a deeper level, undecidability is also the product of our desire to remain vigilant even when—as stated by Ludwig Wittgenstein (1929 in Wood 1999: 107)—we find ourselves 'running up against the limits of language'. In order to shed light on this statement, it is useful to first follow David Wood in his exposition of Wittgenstein's understanding of limits, before turning to the implications that the argument holds for Derrida's understanding of undecidability.

In referencing the limits of language, Wittgenstein (following Heidegger and Søren Kierkegaard) has something other in mind than grappling (beyond knowledge) with different possibilities. Rather, for him, the limit is characterised by 'the *astonishment* that anything exists' (Wittgenstein 1929 in Wood 1999: 107). Astonishment is the outcome of the fact that there are possibilities in the first place, as opposed to nothingness. The reason why the experience of this fact constitutes the limit is because we cannot explain existence—in Wittgenstein's words (in Wood 1999: 107): 'This astonishment cannot be expressed in the form of a question, and there is also no answer to it'. Yet, according to Wood (1999: 107), this astonishment constitutes 'one of *the* fundamental philosophical experiences', even though we cannot resolve our astonishment through means of philosophy, language, or logic. Experiencing the fact of existence serves to confuse rather than to illuminate our understanding, and for this reason Heidegger defines this experience in terms of dread. Wood (109) summarises as follows:

> these *meditations* [on existence] are not ones in which dutiful experience reports back its foreign findings to the throne of philosophy, but rather ones that confuse, disrupt, and disturb any and every demarcation between philosophy and non-philosophy.

Wood further contends that experiencing the fact of existence bears two ethical consequences: firstly, it awakens in us the desire to preserve 'that potential for transformation of *ethos* that comes from recognising that things exist (and might not)' (108); and, secondly, it increases our willingness to be patient, or to stay with the experience of astonishment as such. The first of these consequences will be returned to in the next section, where Nancy's engagement with Heidegger's originary ethics is interrogated in more detail. The second consequence, in which experience is related to ethical praxis, requires further examination in the context of Derrida's understanding of responsible action.

In order to make the connection between experience, responsibility, and undecidability, it is firstly necessary to interrogate Derrida's understanding of experience in more detail. Wood argues that an initial (although incorrect) distinction can be drawn between Derrida's understanding of experience as it comes to the fore in his earlier versus his later works. Under this interpretation, the early Derrida's understanding of experience is linked to (self-)presence, to consciousness, to the purity of speech, and thus to the entire history of metaphysics. In his early works, he uses the term *experience* under erasure[3] in order to distance himself from this notion, which

[3] Under erasure (*sous rapture*) is a typographical expression introduced by Heidegger and employed extensively by Derrida. It involves crossing out words, but in such a manner that they remain legible. It denotes the inadequacy (yet necessity) of using concepts to represent meaning (which is

he defines as 'auto-affection *lived in suppression of difference*' (Derrida 1976: 165). In contrast, the later Derrida no longer uses 'experience' under erasure. He also employs the term quite freely when writing about '"the experience of aporia," "an interminable experience," "the (impossible) experience of death," "the experience of the non-passage," "the experience of mourning," and *even* "the experience of what is called deconstruction"' (Wood 1999: 110).

The reason why Wood contends that rigidly distinguishing between the early and the later Derrida's understanding of experience is faulty, is because Derrida's later understanding of experience is contingent on his earlier understanding. In other words, the fact that the purity of speech is defined by means of suppressing *différance* is not something that we discover through theory and reflection, but something that comes to the fore in the very experience of writing, mourning, or the impossible. Derrida therefore recognises the value in retaining concepts such as *experience* or *subject* (despite the metaphysical baggage that they carry) in order to transform these concepts so that they may signify a radical break with the very tradition that bore them. Wood (113) describes this radical break as 'an abyssal exposure of the *loss* of meaning, of any unity of experience, and of the very idea of the subject.' Whereas for Wittgenstein, the loss of meaning is contingent on experiencing the fact of existence; for Derrida, this loss is premised on the experience of the aneconomic (i.e. that which cannot be assimilated by the history of meaning). Yet, Wood also cautions that these transformed and deconstructed terms may never spin entirely free of their metaphysical heritage—in his words, 'there can no independent "abyssal realm"' (114). He explains the reason for this as follows:

> The experience of the impossible is nothing but the recognition of the impossibility of a certain *closure* of experience. In other words, abyssal thought is directly predicated on the value of *closure*. Without the effort at closure, without the necessary failure of such closure (such determination of meaning, such completeness of identity, etc.), there is no abyss. The abyss is derivative from the experience that it undermines. So abyssal thinking is essentially differentiated from (and hence *dependent on*) that recuperative negation which it refuses (114).

Although risky, Wood contends that Derrida's strategy (i.e. the undermining of presence through the experience of the impossible and the undecidable) is ultimately what gives value to deconstruction. In this vein, Wood (110) claims that:

> deconstruction "itself" and, indeed, the concept of "responsibility,"... are each nothing *other* than experience regained. Deconstruction is, if you like, the experience of experience.

Wood (114) argues that Derrida's strategy does not result in a new graspable truth, but should rather be understood 'as a *way*' (what I have termed *an ethics of living*). Derrida (1999: 74) seems to concur with this interpretation, in writing that 'I don't think deconstruction 'offers' anything *as* deconstruction.... I have never 'proposed' anything, and that is perhaps the essential poverty of my work'. Yet he is also not mute on the question of ethics: he explicitly argues that ethics is about constantly

---

excessive and undecidable). It also highlights our inability to rid ourselves of metaphysical concepts.

facing 'the *aporia* or the *non-way*', which he defines as 'the condition of walking' (73). He further writes that:

> if there was no *aporia* we wouldn't walk, and we wouldn't find our way; path-breaking implies *aporia*. This impossibility to find one's way is the condition of ethics (73).

Wood argues that Derrida's *way* comes to fruition in his notion of responsibility. Wood (1999) traces the origins of responsibility back to Heidegger's reading of Heraclitus, where dwelling is characterised in terms of confronting the unfamiliar (the gods) within the familiar (the hearth). This view of ethos, which implies both an interruption and a deepening, is re-inscribed in Derrida's notion of responsibility, in which:

> The unfamiliar appears as the interruption of experience, in the sense both of experience as interruption and of the interruption of a more domesticated sense of experience (114).

We see echoes of Derrida's (1999: 66) understanding of the *terrible process of undecidability* in the definition of experience as interruption. A responsible decision, as an example of experience, interrupts calculation and constitutes an engagement with alterity, or the other of knowledge. Derrida characterises responsibility (or the interruption by experience) as a trembling, which Wood (1999: 115) describes as 'an experience in which the forces of difference constitutive of any and every identity or presence are activated and acknowledged' to the degree that presence is subverted. Our banal understanding of reality is interrupted and deepened in the moment of the event, which is none other than the encounter with alterity or with the ethical. Derrida (1993) argues that this experience—which is inherently undecidable—also marks the loss of meaning. For this reason the *aporia* of ethics (i.e. the possibility of its impossibility) requires endurance in thinking. We should resist the desire to halt at this *aporia* or to overcome it, and instead focus on the task of translating this sense of loss into '*ways* of going on' (Wood 1999: 117).

Wood however parts ways with Derrida when it comes to his quasi-religious description of our engagement with experience, as expressed in terms such as 'infinite responsibility, absolute singularity, absolute other' (117). According to Wood, this terminology 'constantly threaten[s] to cross the line between a *modal* truth and a renewed mystification, between recursive reminders and impossible prescription' (117). Wood contends that this task of bearing witness to the other by leaving open the door, as demanded by Derrida's notion of unconditional hospitality, is too onerous to be of much use. Derrida concedes to the impossible demands that unconditional hospitality make on the subject (who can also never know this form of hospitality in advance). In this regard, Derrida (1999: 71) writes that, '[there are] risks involved in pure hospitality, if there is such a thing, and I am not sure that there is.' Yet, his commitment to unconditional hospitality—which represents an expression of the aneconomic—is binding. In contrast to Derrida, Wood argues that unconditional hospitality should rather be understood in terms of a willingness to transgress established boundaries. In terms of hospitality, this translates as a willingness to open the door to the stranger. This difference cashes out as follows:

> Responsibility is not quantifiably (or even unquantifiably) *large* and, therefore not a basis for guilt through failure to live up to it. It is rather a recursive modality, an always renewable openness (Wood 1999: 117).

Although it is true that Derrida's tone sometimes verges on the religious, I do not think that he would disagree with Wood's interpretation of the deconstructive enterprise. Indeed, in his 1999 text titled, 'Hospitality, justice and responsibility', which is cited in this section, and which is published alongside Wood's text in Richard Kearney and Mark Dooley's *Questioning Ethics*, Derrida appears sympathetic to Wood's interpretation of deconstruction. This comes poignantly to the fore in his response to the question 'after deconstruction, what is to be done?' (65). He answers as follows:

> Deconstruction… is something which is constantly at work and was at work before what we call 'deconstruction' started, so I cannot periodize. For me there is no 'after deconstruction'… but for what I understand under the name deconstruction, there is no end, no beginning, and no after (65).

Viewed as a recursive modality, deconstruction also does not evoke guilt, but rather necessitates action, even though such action is always problematised by the *aporia*. Derrida also addresses this matter, in speaking about the difference between justice and strategy (law). In this regard, he argues that the demands of justice leave us feeling compelled to act ('I/we must'), but action inevitably requires a process, which is undertaken 'in the name of something which doesn't tolerate the process' (72). In other words, 'I must answer the call of the other… [which is] something which has to be absolute, unconditional and immediate, that is, foreign to process' (72). Yet, the only way in which the experience of the impossible (the demands of justice and responsible action) can find traction in the world is through means of process. Derrida refers to this as the great dilemma, but argues that despite this dilemma, the demands of justice and the necessity of process are *aporetically* tied to each other. It is therefore in the name of responsibility and the 'we must' that 'we have to enter the process, and to analyse and transform infinitely' (73).

This logic, which has been repeated in the discussions presented on Derrida's thought throughout this book, does not sever the experience of alterity from the banal world of experience. Rather it is precisely *because of* the fact that 'there can be no independent "abyssal realm"' (Wood 1999: 114) that Derrida's limit concepts should be understood in terms of an always renewable openness, rather than as representing an impossible negotiation with alterity. In this regard, Derrida often reminds us that justice without law, ethics without action, or—more generally stated—the unconditional without the conditional, would merely remain an empty promise. For Derrida, the impossibility of ethics must therefore serve as the condition for the possibility of ethics. François Raffoul (2008) notes that, in this context, *possibility* and *impossibility* should not be understood as modal opposites; rather, the impossible should be viewed as that which haunts, and thus enables, the possible.

The undecidable—that is, the very impossibility of calculating or systematising morality—is therefore what keeps the pulse of morality alive. Furthermore, it is through the (im)possible experience of grappling with alterity that we are constantly reawakened to the ever-renewed call of ethics. Herein lies both the explanation and the value of Derrida's praxis.

6.3 Nancy: Originary Ethics as Grounds for Praxis

Nancy develops his ethical praxis in a slightly different way to Derrida. Whereas the ethical value of Derrida's work lies in the duty to stay with experience (in terms of neither resisting nor abandoning the *aporia*), Nancy's work is centred on the fact of experience as such. In the previous section, it was noted that Wood identifies one of the ethical consequences of experiencing the astonishment of existence as an awakening of the desire to preserve (and to transform) ethos. It is this line of reasoning that I wish to pick up on in elucidating Nancy's understanding of ethics.

Nancy's (2003a) views on praxis develop at the hand of a close reading of Heidegger's ethics, which he presents in an essay titled 'Originary ethics'. Nancy recognises that Heidegger is not the most obvious ethical philosopher due to the moral taint that stems from his Nazism, and because he explicitly objected to ethics understood as both a discipline and an application. Nevertheless, Nancy argues that Heidegger offers us a profound thinking of ethics. Proof of Heidegger's ethical thought is gleaned from the fact that he 'sought to analyse what it is that constitutes man as the being through whom *being* has its original *sense* (or *ethos*), the choice and conduct of existence' (172). Nancy therefore brackets the common objections that '"Heidegger has a bad morality"' and '"Heidegger has no morality"' (174), and focuses instead on a more fundamental understanding of ethics that comes to the fore in Heidegger's work. As recognised by Nancy, Heidegger's ethics also resonates better with the aims of philosophy as such. Nancy (175) explains as follows:

> It isn't philosophy's job to prescribe norms and values: instead it must think the essence of what makes up *action* [l'agir] as such; it should think, in other words, the essence or sense of what puts action in the position of having to choose norms or values.

The above interpretation of philosophy—which Nancy defines as Heideggerian in tone—resonates with Derrida's understanding of deconstruction. To recall: deconstruction does not propose anything, but incites action understood as a way or non-way. On both counts, the value of philosophy and of ethics lies in action—praxis—rather than in the prescriptive force of normative values or claims. However, the fact that praxis, so understood, cannot be neatly confined to a specific domain also implies that the ethical task is expansive (the *ethics of complexity* is a structural condition for engaging in complex thinking). It is for this reason that Nancy defines ethics in terms of an 'orientation in [Heidegger's] thinking' (173), which necessitates 'a general examination of his thinking' (174). In undertaking this examination, Nancy focuses primarily on the motif of 'original ethics' which is explicated in Heidegger's (1993a) *Letter on Humanism*.

In this text, the issue of praxis also comes starkly to the fore. Already in the first sentence the text announces itself 'as a reflection on *Handeln, action*' (Nancy 2003a: 174). Heidegger explicitly relates the question of humanism (what man is) to man's action, conduct, or praxis. Praxis should be understood as an end in itself, since what is at stake is being itself: it is only through conduct that being forms 'an essential and "active" relation with the proper fact of being' (175). Nancy argues that the proper fact of being should not be understood as a different type of being, but in terms of a relation to *sense*. In this regard, he writes that 'in Dasein it is a matter of giving sense to the fact of being—or, more exactly, in Dasein the very fact of being is one of making sense' (175). Knowledge of sense is equivalent to the action of sense, in that '[t]o be is to make sense' (175), and thinking is furthermore the appropriate mode for action. However, thinking should not be understood in terms of an exclusive or higher form of intellectual conduct. Rather, '[t]hinking (and/or poetry) is… what, in all action, brings into play the sense (of being) without which there would be no action' (175). Otherwise stated, 'sense desires itself as its own action' (176). Being and sense are therefore intimately interlinked, and one cannot posit a logical ordering between the terms. It is not the case that beings first exist and then desire sense. Rather, one must think 'being as the fact of sense and sense as the gift of being' (176). Nancy (177) summarises Heidegger's view of ontology as follows:

> Being, absolutely and rigorously considered as such… is essentially its own "engagement" as the action of sense, therefore: such is the decisive axiom of thinking. From which it follows that ontology is, from the outset, within or beyond itself, being's *conduct of sense* or the conduct of the sense of being, depending on which of these two expressions has the strongest value (the most ethical and least directional value).

Nancy further elaborates in arguing that the value of this view of ontology resides in finitude itself. The notion of finitude does not serve to relate man to some external essence from which he derives his sense or lack of it, but rather denotes the 'non-fixing of such a signification' (178). Otherwise stated, '"Finitude"… means: unaccomplishment as the condition for the accomplishment of action (or for the accomplishment that action *is*) as sense' (178). Returning to Heidegger, Nancy thus argues that the proper fact of being—i.e. 'the facticity of sense that is the ontological fact of existence itself' (179)—takes place in factual experience or the improper mode of everyday existence. Nancy argues that this proximal, yet shrouded, relation between the ontological structure of Being and ontic existence is both the basis of ek-sistence, and what allows the proper dimension of sense to both dissimulate and reveal itself. In Nancy's (181) words: 'Being ek-sists (is) in that it opens being.'

Given this understanding of being, Nancy defines ethics as 'nothing other than existence' (179), where everyday existence is asked—as the proper or original request of its being—to make sense. Nancy moreover argues that it is only on the grounds of this original request that signified content (i.e. all ideas and values) can be both issued and justified. 'The ethics engaged in this way', Nancy (180) states, 'is the bringing to light of making-sense as action requested in the essence of being.' The ethical call thus resides in being, rather than in a beyond. However, since existence is nevertheless structured as a beyond (as an ek-sistence[4]), ethics implies transcendence, understood as:

> the transcendence (of the sense) of being [which] is a transcendence of and for immanence: it is nothing other than the desire/ability of making sense, and this desire/ability *as* making-sense (188).

[4] See footnote 6 in this chapter.

Nancy is careful to stress that desire/ability as making-sense is not a solipsistic or egoistic enterprise, since we share the responsibility towards existence and towards sense. Otherwise stated, 'being's being-responsible towards itself... contain[s] the possibility and the necessity of being-responsible toward others' (180). The reason for this is that ek-sistence, or *Dasein's* mode of being as sense or as the activity of being-open, implies being-with (and hence alterity and difference) as its foundation. In Nancy's (181) words: 'whatever the moral choice, the other is going to be essential to opening, which is essential to sense, which is essential in the action that makes up the essence of being.'

This view of ontology—or ethics defined as ontology—holds the implication that the 'the human is no longer the signified of sense... but its signifier' (183). Here again, we see the emphasis on praxis, since, as Nancy (185) notes:

> What is properly given... is the need to make sense of and in beings as a whole... It is in this sense that humans are responsible for being, or that the Dasein in them is the being-responsible of/for being itself.

Despite this emphasis on praxis, the active mode of being-as-conduct or as thinking should be understood pre-reflexively, as opposed to theoretically. This is because, as Nancy warns, '[t]hinking the origin as *ethos* or conduct isn't the same as representing an originary *ethos*' (184). Related to this issue is the fact that, although thinking as praxis implies conduct, thinking-being should certainly not be understood as intentional-being reflecting on Being. Nancy expresses this view through a short conceptual analysis of the *gift*, which he develops along slightly different lines to Derrida.[5] For Nancy, the gift is impossible or inappropriable as gift, for the reason that nothing is given: '(The) being (of beings) is not a "gift" that it "gives"' (185). Rather, the gift of being is that '[b]eing lets being be' (185), in the sense that it opens onto sense, but 'not *as* a sense or as an appropriable horizon of signification' (185; my italics). The gift of being thus lies between passivity and activity (what Nancy denotes by *letting-be*), and Nancy (186)—citing Heidegger (1993a: 235)—defines the appropriate gesture for the gift of being as that of touching:

> existence touches itself; in other words, it "moves" itself, sets itself moving outside of itself and affects itself with its own *ek*- [[6]] ... "Nearness" and "touching" evoke what we would have to call the intimate distance according to which "being" is related to "the essence of man," in other words, according to which "being itself is the relation", Being *is* the relation of existence to itself as the action of sense (Nancy 2003a: 186).

In this description of being, we again see that the possibility of ethics resides in the conduct of existence itself, which Nancy defines as the action of sense. Ethics should thus not be understood as the act of acquiring sense, but as 'the ontology of ontology itself' (Nancy 2003a: 187). This is because, as Nancy argues, acquisition implies signified content, which would serve to define ethics in terms of a mere

---

[5] See Sect. 3.4.2.1.

[6] The prefix ex- (Latin) or ek- (Greek) means 'out of' or 'outside of'. It is derived from *exclosure* (an area from which unwanted animals are excluded) and *exurbia* (the area outside of the city).

disciplinary ordering of moral significations or values, and thereby foreclose the opening of ethics as such. Ethics, defined in terms of making-sense, is thus prior to any disciplinary divisions, 'just as the conduct of existence is prior to any determinations of signification' (188).

Given this short exposition, we are now in a better position to provide a Nancean interpretation of Wood's claim that a possible ethical response to existence as such is an awakening of the desire to preserve and transform ethos. Two points are important with regard to this formulation. The first concerns our understanding of ethos, whereas the second is related to what is meant by preserving and transforming ethos. Nancy notes that we should understand ethos as an abode, which in turn should be understood as the activity of residing (rather than passively as 'residence'). The abode or ethos is therefore principally a conduct or originary ethics. This involves 'thinking *ethos* as the conduct of/according to the truth of being' (188). However, and as concerns the second point, thinking should not be thought of as an outcome or signified content, because this again implies mastery of thought. As such, it would destroy both making-sense and dignity, which Nancy (189–190) describes as follows:

> thinking is its own result… It posits and it posits itself actively, which is also to say that it obligates itself to encounter human dignity insofar as the latter is incommensurable with a fixing of signification and a filling out of sense… Dignity is possible only if it measures up to finitude, and finitude, as will now be clear, means the condition of a mode of being whose sense *makes-sense* as a ground and a truth.

The preservation and transformation of ethos should therefore not be thought of in terms of the conscious activity of the knowing subject on predefined being. Rather, it is the outcome of praxis, defined as being-engaged in thinking. However, opening ourselves to the gift (or to making-sense as such) also constitutes an opening to the possibility of evil, defined as 'the possibility of not receiving the gift as a gift' (191). Nancy does not expound further on the nature of evil, except to note that it is possible 'as the "rage" that precipitates being into the nothing that it is also' (191). As with being, evil should not be thought of substantively, since, 'any determination of evil would lead us away from the necessity of thinking the possibility of evil as a possibility of ek-sistence' (192).

What the reference to evil denotes is that sense is not inevitable. Although Nancy does not give a further indication of why this is the case in the text 'Originary Ethics', more information can be gleaned in following Nancy in his deconstruction of the notion of freedom that informs Kant's categorical imperative.[7] As explained by Benjamin Hutchens (2005), for Nancy freedom does not manifest in our choice for or against the imperative as Kant would have it, but precedes the very concept of conscious choice. Freedom, in other words, is the very condition of the imperative that, as an imperative, constitutes the will or duty to sense (understood as the condition of existence). Freedom also implies a radical notion of evil, in that evil is the freedom to will against existence and sense; and, hence, to will against itself.

[7] This argument is drawn from Woermann (2013: 281–282).

Freedom, defined in terms of the will or duty to sense, confronts thinking as a certain non-thinking. In other words, freedom is that which allows us to think by revealing the finitude and elusiveness of thought, which, as explained above, are the enabling conditions of thought. In the text 'Concealed thinking', Nancy (2003b: 34) explicitly relates thinking to freedom, in writing that:

> There is no thinking, no articulation of sense, that doesn't have something of the uncompleteable about it, that doesn't exceed sense, like an intimation, a binding, implacable obligation, logical as much as ethical, to conceal itself as thinking in the very act of thinking "in order," if you'll allow me to risk the phrase, to be thinking ("in order to make sense" and "in order to free itself"...)

This freedom generated by thought reveals itself in a Bataillian burst; or a surprise—the surprise of thought, where thought surprises itself (recall again Wittgenstein's astonishment at the sheer fact of existence). Responsibility, which emanates from thinking defined as making-sense, constitutes thinking as a duty towards sense. In 'Originary ethics', Nancy (2003a: 186) writes that:

> nothing is more ordinary than the call... to the "sense of existence," and nothing is rarer than responding to this call in a fitting ("responsible") way, in other words, without being deceived by a "sense" supposedly given to existence, as from within or beyond it, instead of confining ourselves to the making-sense of ek-sisting.

In order to make-sense of ek-sisting, responsibility must therefore interrupt, or differ, from itself. Indeed, Nancy (in Lacoue-Labarthe and Nancy 1997: 49) writes that in the imperative (or duty) to make-sense 'reason... speaks to itself, it addresses itself to itself, but it does not hear itself there: it cannot call to account the theory of its freedom. As a result, it separates itself from itself.' According to James Gilbert-Walsh (2000: 46), it is this interruption (the simultaneous presentation and withdrawal of the imperative) that highlights the inadequacy of understanding, and that invites the other's reciprocal rejoinder. This again affirms the fact that 'responsibility only ever takes place as a responsibility with and toward others' (Nancy 2003a: 191). 'And', Gilbert-Walsh (2000: 46) asks:

> is not the possibility of such a rejoinder the very possibility of a continued *accomplissement* of interruption, of *praxis*, of 'that echo which is not a repetition'? Perhaps this is the force behind Nancy's imperative: '[The] work... must be offered up for communication' (CD, 181/73).

For Nancy, thinking or making-sense is always an activity that is related to some form of alterity. This otherness, which can never be appropriated by a body of philosophy, nevertheless propels us to communication and to doing philosophy, understood in terms of the preservation and transformation of ethos. The manner in which Nancy's version of praxis is thus enacted with others in the world is through communication. The implication is that '[c]oncealed thinking is identical to communication, and this identity itself is the night of not-knowing'[8] (Nancy 2003b: 45).

---

[8] See Sect. 3.5.2.

6.4 The Status of Philosophy

The above analyses of Derrida's and Nancy's views on ethics serve to draw attention to the similarities and the dissimilarities between them. At this juncture, it should be clear that both these thinkers seek to recast our understanding of ethics and philosophy in terms of praxis rather than disciplinary knowledge. However, their interventions are pitched at different levels.[9]

Derrida seeks to preserve, but radically transform, our traditional concepts such as *experience* and *subject*. He works from within the history of metaphysics (the *symbolic-ideological order*) in order to break it open and transform it through means of a deconstructive intervention. In so doing, he focuses our attention on, or creates an opening for, the experience of difference and alterity. Moreover, and as is clear from the text, 'The right to philosophy from the cosmopolitan point of view', Derrida (2002c) conceives of philosophy as the proper activity for bringing about such an opening.

In this text, Derrida interrogates not only the nature of philosophy, but also the cosmopolitical right to philosophy. As concerns the former, Derrida again employs his strategy of working within the history of meaning in order to transform it. This comes to the fore in his question regarding whether:

> the unlimited extension and the reaffirmation of a right to philosophy should not both *take into account and de-limit* the assignation of philosophy to its Greco-European origin or memory (9).

Derrida neither rejects nor affirms the history of philosophy, but instead tries to move beyond the oppositions that structure the traditional account of philosophy in order to transform philosophy and reassert the impossibility of its closure. In this regard he writes that:

> There are other ways for philosophy than those of appropriation as expropriation… Not only are there other ways for philosophy, but philosophy, if there is such a thing, is the other way (10).

Next, Derrida raises three points for discussion regarding the future direction of philosophy, and how the right to philosophy could be extended to all. Going much too fast, these points amount to the deconstruction of philosophical hegemonies (and the exclusive right of access); to engendering the irreducible autonomy of philosophical thought (to the extent that thought cannot yield to *isms* or be made subservient to history); and, to re-instantiating the urgent need for philosophy as the means through which to critically evaluate the substantive philosophies of our economic and political regimes.

In discussing the nature of philosophy, Derrida argues that 'philosophy has never been the unfolding responsibility for a unique, originary assignation linked to a unique language or the place of a sole people' (10). Rather, philosophy 'has always been bastard, hybrid, grafted, multilinear, and polyglot' (10). Given this description,

---

[9] Part of this analysis is informed by the arguments posited in Woermann (2013: 281–283).

it could be argued that Derrida's view of philosophy becomes a vehicle through which his experiential view of praxis can come to fruition. This is because, in doing philosophy, one puts the body of the idiom of philosophy 'into operation, each time in an original way and in a nonfinite multiplicity of idioms' (12).

In contrast to Derrida, Nancy's ethical praxis is not defined in terms of deconstructing the symbolic-ideological order (what he terms the *operative community*) in order to facilitate the experience of otherness. Rather, praxis is encapsulated in our duty to respond to sense. Originary ethics therefore does not supersede the text, but is itself instantiated in, and as, the existence of the text. It is for this reason that Nancy (2003a: 187) defines ethics as 'the ontology of ontology itself.'

In light of this comparison, Derrida's position may seem more conservative, in that he still works with the history of metaphysics in order to demonstrate the impossibility of its closure. Nancy, in contrast, provides us with an ante-philosophical understanding of ethics, which not only serves to strip duty of its traditional understanding, but also seeks to avoid (as opposed to deconstruct) the abstractions upon which the disciplines of ethics and philosophy are based. However, it should be remembered that, for Derrida, the impossibility of the closure of meaning is premised on a radical alterity, which renders all signified content (including ethics) as *a priori* ruined. On this reading, ontology itself (i.e. the existence of the text) is threatened by the aneconomic force of *différance*.

Nancy is emphatic in his opinion that ethics and philosophy, indeed the humanities as such, cannot be reduced to an 'assignable value' or a 'measure of action regulated by a particular given' (178). In this regard, he identifies the inadequacy of humanism as lying in the assumption that we can know being and fix sense. He instead argues that '*Humanitas* needs to be measured against this measurelessness of action, or rather, action itself as the absolute measure' (178). The fact that Nancy's conception of thinking, communication, or praxis denies any form of closure is also the enabling condition of these activities. Citing Heidegger (1993b: 294), Nancy (2003a: 189) notes that if 'originary ethics were to provide "maxims that could be reckoned up unequivocally," it "would deny to existence nothing less than the very *possibility of acting*."' True thinking, therefore:

> rejects abstraction and conceptualisation as these are recognized by understanding. Thinking does not produce the operators of knowledge; it undergoes an experience and lets the experience inscribe itself (Nancy 2003c: 247).

Derrida, in contrast to Nancy, accepts the fact that the humanities are largely built on the assumption that being and sense can be fixed. Yet, he argues that although philosophy is a designated discipline within the humanities (with its own signified content), philosophy is also 'the discipline that claims to think, elaborate, and criticize the axiomatic of the "humanities," particularly the problem of humanism or the presumed universalism of the "humanities"' (Derrida 2002c: 1). The critical role of philosophy is premised on the recognition that our conceptual schemas (in this case, 'the axiomatic of the "humanities"') cannot exhaust meaning. It is this realisation that serves as impetus for deconstruction and for thought as such. As with Nancy, Derrida therefore recognises that thinking is designated by non-closure. It is not signified content, but the ever-renewed obligation to tackle that which does not yield to reified meaning, that grounds philosophy.

In summary, the positions of both Nancy and Derrida affirm the view that moral theory (which is necessarily based on abstraction and conceptualisation) and its application can never coincide with ethical duty, and hence moral theory ceases to be what it purports to be—i.e. moral in nature. It is only by virtue of thinking that we—as limited and finite beings—can experience our ethical duty, which manifests as an experience of alterity (Derrida) or the will to sense (Nancy). Both these experiences are premised on the impossibility of closure. Ethics therefore has 'a failure of achievement… as its very propriety, absolutely and unconditionally' (Nancy 2003b: 32).

In practical terms, our engagement with the *aporia* (Derrida) or our will to sense (Nancy) manifests as the continually renewed engagement with thought in every activity of our daily life (including the activity of *doing* philosophy). Applying Derrida's and Nancy's work to moral philosophy therefore offers us a perspective that conflates ethics' (indeed, philosophy's) nature and function. Whilst both Derrida and Nancy concern themselves with a level of engagement that precedes the fixing of sense, Derrida's deconstructive philosophy additionally constitutes a critical engagement with the text or reified sense in order to demonstrate the impossibility of its closure. On both counts, ethics (understood as something other than moral values) becomes performative. In this regard, it is useful to note Nancy's description of praxis as 'the endless transformation of the subject of sense in itself: a sense that is nothing other than its communication—and, by the same token, its concealment' (47).

Thus, despite the differences between Derrida's and Nancy's views on praxis, their views share three important implications, all of which stem from defining ethics as a recursive praxis, rather than as a universally-binding moral theory. Firstly, if philosophical concepts have no final grounding, then the value of philosophical argumentation lies not in the pedagogical or prescriptive nature of philosophy, but in the constitutive nature of philosophical thought. Secondly, the attempt to divorce morality from the metaphysical project, and, instead, to describe morality in immanent terms, means that morality is progressively viewed in terms of the performative force of *doing* philosophy, rather than in terms of the locutionary force of substantive moral statements. In other words, philosophical form, rather than content, becomes important. By taking the first two points together, and by employing a deconstructive gesture, one could say that all philosophy becomes a type of applied ethics (but not in the traditional understanding of the term), which serves to problematise the distinction between the nature and function of philosophy. Thirdly, the status of philosophy is challenged, since philosophy becomes one type of praxis amongst many, rather than the definitive voice on praxis. In this regard, recall that Nancy insists that thinking should not be viewed as an exclusive enterprise, reserved for the learned; and, that Derrida argues that all people should enjoy a cosmopolitical right to philosophy, which necessitates the deconstruction of philosophical hegemonies. Yet, the practise of philosophy still remains a vital tool for sense-making. This is because philosophical competence aids us in successfully engaging in the logic of the *aporia*, in which identity and meaning is irrevocably tied to alterity and non-meaning; or, in exploring positions like Nancy's, that forego the *aporia* in order to understand philosophy in terms of the duty to think our concealed thoughts.

6.5 Ethical Praxis and Complexity

Both Derrida and Nancy foreground thinking (in Derrida's terms, *decision-making*), as opposed to the justification and application of normative principles, as the grounds for ethical praxis. In the foregoing discussions, it was demonstrated how such a move serves as the very condition of thought and action, and therefore of ethics itself. In this section however, I wish to shift the analysis from the conceptual level to exploring the implications that the analysis holds for the ethics of living. Specifically, I argue that Derrida's and Nancy's views of ethics necessitate modesty, openness, and a sensitivity for complexity.

6.5.1 Modesty and Openness

To define ethics in terms of the impossibility of closure serves to draw attention to what I defined in the introduction as the *ethics of complexity*, which simply amounts to the fact that no complete description of complex phenomena is possible (in Nancy's terms, we cannot fix sense). As a result, normative concerns are always present, in terms of both the process of making-sense (Nancy), and the evaluation and transformation of the products of sense-making (Derrida). The ethics of complexity thus focuses our attention on the fact that, ethically-speaking, we are always in trouble (because meaning is always incomplete), but that our woes are the lifeblood of ethics and of being as such. Despite the fact that it is only on the basis of an originary ethics or an ethics of alterity that beings can both assign and justify ideas or values (Nancy 2003a), the ethical positions espoused here do not allow for substantive ethical systems.

The question that now arises is whether we can move beyond this position in order to say something more about the *complexity of ethics,* other than to note that competing interpretations or models of the good life are possible, and that these interpretations cannot say anything about ethics understood as either the grounds of being (Nancy) or the opening onto alterity (Derrida). In other words, the question is whether recognising the *complexity of ethics* can, in any way, alter our understanding and application of ethical models. In the article, 'The ethics of complexity and the complexity of ethics' (Woermann and Cilliers 2012), we argue that recognising ethical complexity can help in the development of a meta-ethical position, which can serve to highlight important considerations that underscore the ethical strategies that we employ when engaging in the particularities of situations.[10]

As a starting point, Cilliers and I turn to perhaps the most famous example of a meta-ethical position in the history of moral philosophy, namely Kant's (1993) categorical imperative. The categorical imperative is a substantively empty rule, in that

[10] The discussion on the meta-ethical position presented here is taken from the article by Woermann and Cilliers (2012: 451–452).

it cannot generate contingent ethical principles, but can merely act as a yardstick for evaluating the morality of principles that already exist. This is because Kant wants his moral rule to be categorically applicable and, hence, universally valid. However, the only rule that conforms to this criterion is a purely abstract and formal rule, which says 'always follow only universal rules'; or, otherwise stated, 'always follow only rules that you will want all other people to follow'. Thus, by combining a purely formal rule with the notion of universability, Kant can generate a formulation that actually does say something about ethics, namely that if certain contingent principles are universalisable, then the principles are deemed morally correct. Therefore, although the categorical imperative cannot indicate which principles are good, right, and deserving of respect, it does provide a strategy for evaluating our contingent principles. As such, one can argue that Kant's categorical imperative urges us to adopt a certain strategy when undertaking moral considerations (Preiser and Cilliers 2010).

Next, we attempt to apply the same logic that Kant uses to the *ethics of complexity* in order to say something about the *complexity of ethics* (in other words in order to develop a meta-ethical position). Ethical complexity implies that all knowledge is contingent, subject to revision, and therefore irreducibly provisional. Following Kantian logic, we capture the gist of the above argument in the following imperative: 'When acting, always remain cognisant of other ways of acting'. Our meta-ethical position thus constitutes a *provisional imperative* (Preiser and Cilliers 2010).

Note that on one reading, the idea of a provisional imperative is a contradiction in terms, since the logic of an imperative is absolute: either one chooses to follow the imperative or one ignores it. The idea of a *provisional* imperative seems to suggest that the imperative itself is subject to change, and in this regard we seem to be advocating an impossible position. This is, to a large extent, exactly the point: we cannot do away with moral imperatives, but, if we take complexity seriously, we should also realise that our imperatives do not live up to the demands of ethics (as previously stated, moral theory and its application cannot coincide with ethical duty). Thus the provisional imperative stipulates that we must be guided by the imperative, whilst simultaneously acknowledging the exclusionary nature of all imperatives.

In terms of the actual content of the imperative, it should be noted that—unlike the Kantian imperative—which tells us something about the rules for action, the provisional imperative says something about our state of mind or *attitude* when choosing rules for action. Again: it is impossible to say that 'When acting, always choose rules that admit to the possibility of other rules', since the logic of rules (as with the logic of imperatives) is absolute. In this regard, Derrida (1988: 116) notes that:

> Every concept that lays claim to any rigor whatsoever implies the alternative of "all or nothing"... Even the concept of "difference to degree," the concept of relativity is, qua concept, determined according to the logic of all or nothing, of yes or no: differences of degree *or* no differences of degree. It is impossible or illegitimate to form a *philosophical concept* outside the logic of all or nothing.

In the above citation, Derrida is pointing to a structural condition of all concepts. We cannot do other than to model and to exclude. Yet, what the provisional imperative tells us is that when we act, we must be cognisant of this logic.

In this regard, we argue that it makes a difference—and moreover, an ethical difference—whether one exercises this awareness. This is because if we remain open to other ways of modelling and other ways of being, we are more likely to practise a self-critical rationality, to respect diversity, to be willing to revise our models, and to guard against the naturalisation of these models. The provisional imperative therefore provides us with a strategy for remaining open to complexity at the same time that we reduce complexity through our decisions and actions.

The attitude of modesty that underscores the provisional imperative resonates with the bearing of language that Heidegger describes in his *Letter on Humanism*, and which Nancy (2003a: 193) translates as follows:

> the bearing of language... expresses what are, properly speaking, its only maxims, the maxims of "bearing" itself: "rigor of meditation, carefulness in saying, frugality with words." These three maxims propose no values. Nor could they be used simply to measure the "ethicity" of any given discourse... These three maxims are merely the maxim of the measure of language in its relation with the unmeasurable character of making-sense.

Applying the above description of language to the provisional imperative yields the insight that the attitude of modesty that is evoked in the application of this imperative should not be read as an act of virtue, but merely as a recognition of the limitations that define our actions. In speaking of Heidegger's conception of language, Nancy (194) summarises as follows:

> "Bringing to language"... means entrusting the acts of language, as all acts, to the conduct of sense, to the finitude of being, in other words, to the ek-sistence in which "man infinitely exceeds man."

Language is the medium through which the preservation and transformation of ethos takes place; and, in elucidating the maxims of language, Heidegger (and Nancy) serves to draw attention to both the limitations of meaning, and the excesses that it gives rise to. What the three maxims of language and the provisional imperative thus share is that they are designed to achieve the opposite of traditional moral theory: instead of fixing the limits of moral conduct (in terms of values, norms, and principles), the maxims and the provisional imperative challenge us to transgress these limits, by focusing our attention on the inherent limitations of signified content expressed in language and in action.

Von Foerster (1990) shares a similar view regarding the bearing of language. Following Wittgenstein, he argues that ethical reward resides in practice itself, and consequently each of us should try '*to master the use of my language so that ethics is implicit*.' By this he means that one should attempt:

> to let language and action ride on an underground river of ethics, and to see to it that one is not thrown off, so that ethics does not become explicit, and so that language does not degenerate into moralization.

To my mind, to be aware of this undercurrent of ethics means to contend with the excesses or the manner in which "man infinitely exceeds man" (to again quote Heidegger), which is facilitated by practising the provisional imperative. Von Foerster further reminds us that this can only be achieved when 'my language reaches out for the other', which he characterises as 'the root of conscience… where ethics invisibly manifests itself through dialogue.'

6.5.2 Locking Eyes with the Stranger

Although it is impossible to understand the infinite ways in which finite meaning exceeds itself—in Nancy's (2003a: 182) words, although 'no norm or value can be determined on the fundamental level, where what is at issue is valueless value'—it certainly does not mean that ethical praxis should lead to 'indifferentism' (defined in terms of either subjectivism or moral relativism). In order to avoid this conclusion, Nancy argues that the truth of ek-sistence, and the imperative that it gives rise to, namely '*respect existence*' (183), should serve to quasi-orientate action. This imperative provides no substantive content; however it does require that we attempt to make sense of existence *as* existence, without reducing it to a specific content (such as, for example, *respect for life*). Yet, in, for example, trying to determine the content of what it means to respect life, we invariably engage the imperative of respecting existence, to the extent that we involve ourselves in praxis as making-sense. Nancy (183) argues that:

> all the problems being raised today… bring to light the necessity of heading back toward an ontology of action: not so that they can be resolved once and for all, but so that we can apprehend the absolute making-sense of the action that puts itself in the position of having, for example, to decide what a "human life" is—without ever having the ability to fix this *being* as a given that has been acquired once and for all.

Thus the virtue of philosophising and thinking lies not in the outcome of thought (the substantive sense that thought gives rise to), but in the awareness of ek-sistence, as an infinite and non-appropriable exposure. And, as argued above, this exposure constitutes 'a reaching out for the other' in language (von Foerster 1990). Indeed, to respect existence means to respect our being-together and being-responsible toward others. As highlighted by the positions explored in this work, there can be no *cogito ergo sum* without the *cogito ergo sumus*. This is because the proper function of language, and the medium of thought, is dialogics—in von Foerster's words: "'It needs two to language'". In order to stress the relational aspect of being that dialogics introduces (which he describes as the hand-maiden of ethics), von Foerster cites the last few lines of Martin Buber's (2001) book, titled *Das Problem der Menschen,* in which Buber writes that:

> We may come closer to answering the question "What is human?" when we come to understand him as the being in whose dialogic, in his mutually present two-getherness, the encounter of the one with the other is realized and recognized at all times.

In order to try to give more content to what this might mean in practice, I wish to return once more to the experience of strangeness (which is nothing other than the experience of ek-sistence). For this analysis, I shall not draw on personal anecdotes, or the thinking of philosophers, but instead turn my attention to a recent essay written by award-winning journalist, Jonny Steinberg (2015), titled 'Why I'm moving back to South Africa'. In my interpretation, this essay constitutes an attempt to describe praxis or the interruption of reified sense, which is nothing other than the experience of ek-sistence (or the being-responsible towards/with others).

Steinberg commences by offering a detailed description of the gentle life that he leads as a resident of Port Meadow and a tenured academic at Oxford University. He then goes on to state that he has resigned his job, and is moving back to Johannesburg, which he defines as 'a city that heaves with umbrage'. Although, he is quick to acknowledge that he will not be living a life of hardship in Johannesburg, he also states that the move goes against the grain of his family values. In this regard, he writes that 'I am a Jew. My kind tends to sniff out trouble generations in advance. We like the foundations beneath our feet to run deep.' Why then the decision to move?

The description that he offers captures something of the nature of the insights offered in this work. It speaks of experiences of difference, strangeness, and even danger, and shows how these experiences facilitate a silent communication, language, or an experience in making-sense, whilst simultaneously interrupting the identity of a phenomenological ego. I cite his attempt at explanation at length:

> When I lock eyes with a stranger on Johannesburg's streets, there is a flicker, a flash communication, so fast it is invisible, yet so laden that no words might describe it. This stranger may be a man in a coat and tie, or a woman who wears the cotton uniform of a maid, or a construction worker stripped to the waist. Whoever he is, he clocks me as I pass, and reads me and my parents and my grandparents; and I, too, conjure, in an instant, the past from which he came. As we brush shoulders the world we share rumbles around us, its echoes resounding through generations. He may look at me with resentment, or longing, or with the twistedness that comes with hating; he may catch me smiling to myself and grin. I am left with a feeling, both sweet and sore, that I am not in control of who I am. I am defined by the eyes that see me on the street. I cannot escape them. I cannot change what they see. We may one day fight one another or even kill one another, yet our souls are entwined because we have made another.

To my mind, this description constitutes a very articulate exposition of the moments in which one is confronted with the astonishment of the sheer fact of existence (the being-with), and the attempts at making-sense that take place in these moments. This description should not however be read as the paradigmatic case of making-sense (there can be no textbook case), but merely as an illustration of what concealed thinking (which is nothing other than implicit ethics) could mean in practical terms.

One of the strengths of this description is that it demonstrates the non-elitist nature of thought, where, to re-cite Nancy (2003a: 175), '[t]hinking (and/or poetry) is… what, in all action, brings into play the sense (of being) without which there would be no action.' Yet, despite being non-elitist, Steinberg (2015) is quite evidently of the opinion that not all contexts stimulate thinking in the same manner. According to him, the astonishing moments cannot be experienced in Port Meadow for the reason that not enough is at stake. There, the gentleness of daily life covers over the imperative to think, to communicate, and—in so doing—to respect existence. Consequently, he argues, one is left with the impression that 'the people who pass are wafer thin… [and] I am in essence alone'. It is debatable whether Steinberg's impressions in this regard are correct. I would argue that thinking is non-elitist in that it can neither be the right of the learned, nor the right of those who inhabit a certain place (even when the value of place may reside in the fact that it is not built on European memory). However, in the same vein that one could argue that philosophical skill is a virtue given the endurance that thinking demands, one could also argue that certain contexts may stimulate greater awareness of the need to preserve and transform one's dwelling, and is so doing, stimulate ethical praxis.

Moving on in the essay, and still staying with the reasons for his move, Steinberg reflects on his life as a journalist. He notes that, over the past 15 years, he has attempted to document the history of South Africans whose experiences are quite unlike his own. Amongst his subjects are a prison gangster and a refugee. He candidly admits that the reason for 'riding on the coattails of their histories' is to try to make sense of 'my inscrutable country'. Although he recognises that he can never assimilate their foreignness, he nonetheless states that he tries to imagine, as fiercely as he can, how the world appears to these other lives that are so different to his own. In other words, he enacts the provisional imperative to the extent that he remains aware of other ways of being, whilst forging a very specific, and a very different, life path of his own.

He notes that the greatest moment of clarity into the lives of others comes from paying attention to the moments of decision-making. He further argues that '[t]he more puzzling the decision, the further one must reach in order to understand, the better. If I can get an inkling of why a person decides, I can begin, if just fleetingly, to inhabit him'. Drawing from Derrida's exposition on undecidability and decision-making, we can argue that understanding the stranger entails taking seriously the choices she makes, given the over-determinations and the constraints within which she operates; and that, because these choices take place beyond codified knowledge, they tell us something about a subject who is more than the product of calculation. In von Foerster's (1990) words: 'we are under no compulsion, not even under logic when we decide upon in principle undecidable questions… The complement to necessity is not chance, it is choice! We can choose who we wish to become when we have decided on in principle undecidable questions.' As such, our choices and decisions also give others insight into who we are.

Yet, Steinberg is correct is arguing that any habitation of the other is fleeting. This is because, if the subject's actions lie beyond calculation, and if the subject is constituted through praxis, then the other remains unknowable or undecidable (both to herself and to others). The bulk of Steinberg's (2015) essay consists of a description of the life of one such other that has crossed his path, namely Asad Abdullahi, a refugee from Somalia. Steinberg relates the story of how Asad had come to find himself in South Africa; the daily challenges with which he is faced; and, his motivations for repeatedly 'courting death', given the xenophobic attitudes of South Africans towards African foreigners. Steinberg defines this anecdote *as* his explanation for going home, and he concludes the essay with a description that, to my mind, captures something of the ethics of living:

> I am quite unlike Asad... Yet I have imagined the world through Asad's eyes as fiercely as I can, and have thus been under the skin of a human being I am not. The importance of this experience is ineffable. It is to watch oneself from a distance and imbibe the contingency of who one is and what one feels. This is a secular incarnation of the oldest religious experience.
>
> That is what going home means for me. It is to stand outside myself and watch my bourgeois life prodded and pushed and buffeted around by lives quite unlike my own. It is to surrender myself to a world so much bigger than I am and to the destiny of a nation I cannot control. In this surrender is an expansion, a flowering, of what it means to be alive.

6.6 Conclusion

At this juncture, it should be clear to the reader that the ethics of living cannot be delineated in terms of disciplinary boundaries. In other words, the ethics of living does not merely constitute an alternative to our traditional moral theories (as described in the introduction to this chapter), but challenges our understanding of ethics, ontology, and epistemology as such. Specifically, what the complexity and the philosophical positions that were interrogated in this study imply, is that we cannot sever the question of the good from either the question of knowledge or the question of identity. For the sake of academic expediency, these three questions were treated separately in different chapters of this work, but even given these artificial demarcations, the analyses kept bleeding into one another. During the course of the preceding discussions, explanations were given for the imbrication of these three central philosophical problematics, and I cannot summarily reproduce these explanations here without infringing on the particularities of the positions and arguments that were referenced.

One insight that does however undergird all the arguments presented in this work is that the clear delineation between ontology, epistemology, and ethics can only be defended on the basis of a rigid distinction between the economic and aneconomic dimensions of meaning, in which the aneconomic never threatens the order of economic meaning. The fact that the aneconomic dimension resides in the heart of the economy (or, in complexity terms, the fact that there is no central organising principle) denotes the impossibility of closure, and hence the impossibility of rigorously delineating our three primary philosophical questions as distinct and complete fields of inquiry.

Ethical considerations however permeate the analyses presented in this work. Note that during the course of this work, the meaning of ethics has undergone a transformation, and the question of ethics can no longer be unproblematically translated as the question of the good. In the analyses provided, ethics has been described in several different ways, which are briefly summarised below.

The main insight to have come out of the complexity literature is that ethics is a structural condition of any engagement with complex phenomena. The reason for this is that, if we cannot know the world completely, and if we are—in part—the product of our interactions in complex systems, then neither the subject-object nor the fact-value distinctions hold. To explain further (and with specific reference to the subject-object dichotomy), the subject is constituted in a network of relations with others and in the world. According to Morin, this commits us to a co-constructivist position, wherein we form the world and others, in as much as the world and others form us. Moreover (and with specific reference to the fact-value dichotomy), every description of the world or of others necessarily implies a normative dimension. This is because modelling or drawing boundaries is the product of both physical properties and mental constructions. Our mental constructions also cannot be justified in terms of *a priori* obligations, or with reference to a predefined purpose, but are instead influenced by, and defended in, context.

The deconstruction of both the subject-object and the fact-value dichotomies serve as a basis for Derrida's *aporetic* logic, in which we are tasked with thinking together the realm of order (reified sense or fact) and the realm of alterity (ethics). Derrida concedes that, logically speaking, the economic and aneconomic dimensions are mutually exclusive, and yet they are irrevocably tied to each other to the extent that the realm of the impossible enables, but also challenges, the possible. The non-closure of meaning drives meaning forward. Deconstruction, defined in terms of the continuous transformation of meaning, is undertaken in the name of what Derrida calls *ethical testimony*. He writes that ethical testimony, which implies 'faith or promise', is absolutely central to deconstruction for the reason that '[i]t is only by reference to the possibility of testimony that deconstruction can begin to ask questions concerning knowledge and meaning' (Derrida 1999: 82). On this take, ethics therefore implies vigilance over the non-closure of meaning, which we experience as an openness to alterity. This is something that we become acutely aware of when undergoing the terrible experience of undecidablity that precedes decision-making.

Nancy describes ethics as the experience of ontology, in that the non-closure of meaning *is* the defining feature of our ontology. Whereas in Derrida's work, the non-closure of ontology both facilitates and destroys meaning; Nancy argues that the pre-theoretical recognition of ontological non-closure compels us to the imperative to make-sense. In so doing, we preserve and transform ethos through praxis (defined as the action of thinking or of sense).

To define the ethics of living as other than moral theory or signified content is not to deny the pragmatic value of moral theory. Indeed, Wood (1999: 105) argues that '[w]e owe to concepts like "justice," "rights," "duty," "virtue," "good," "responsibility," and "obligation" our very capacity for ethical judgement.' And yet, he also warns that:

> the work of clarifying and codifying the scope and significance of these terms is the source of another danger—the calculation of our responsibility, in which the ethical as an openness to the incalculable is extinguished (105).

Drawing attention to the ethics of living counteracts this danger, to the extent that the focus of such an ethics resides in a meditation on 'the ethical as an openness to the incalculable'. As such, the ethics of living is nothing other than an engagement with the intelligence of complexity, the purpose of which Morin (2007: 29) describes as follows:

> The intelligence of complexity, isn't it to explore the field of possibilities, without restricting it with what is formally probable? Doesn't it invite us to reform, even to revolutionize?

In order to fully respond to the challenge issued by Morin, the following final chapter is dedicated to summarising the implications that this study holds for our understanding of philosophical complexity. The hope is that in so doing, the study will not only present a challenge to our traditional understanding of epistemology, ontology, and ethics, but will also forward our thinking in the field of philosophical complexity.

References

Buber, M. 2001. *Das Problem des Menschen.* Gütersloh: Gütersloher Verlagshaus.

Cilliers, P. 2005. Complexity, deconstruction and relativism. *Theory, Culture & Society* 22(5): 255–267.

Derrida, J. 1976. *Of grammatology.* Trans. G. Spivak. Baltimore: John Hopkins University Press.

———. 1978. Structure, sign, and play in the discourse of the human sciences. In *Writing and difference*, ed. A. Bass, 278–294. Chicago: University of Chicago Press.

———. 1988. Afterword. Trans. S. Weber. In *Limited Inc.,* ed. G. Graff, 111–160. Evanston: Northern Western University Press.

———. 1993. *Aporias.* Stanford: Stanford University Press.

———. 1999. Hospitality, justice and responsibility: A dialogue with Jacques Derrida. In *Questioning ethics: contemporary debates in philosophy*, ed. R. Kearney and M. Dooley, 65–83. London: Routledge.

———. 2002a. As if it were possible, "within such limits"... In *Negotiations: Interventions and interviews, 1971–2001,* ed. and trans. E. Rottenberg, 343–370. Stanford: Stanford University Press.

———. 2002b. Ethics and politics today. In *Negotiations: Interventions and interviews, 1971–2001,* ed. and trans. E. Rottenberg, 295–314. Stanford: Stanford University Press.

———. 2002c. The right to philosophy from a cosmopolitical point of view: The example of an international institution. In *Ethics, institutions, and the right to philosophy,* ed. and trans. P.P. Trifonas, 1–18. Maryland: Rowman and Littlefield Publishers.

———. 2005. The principle of hospitality. Trans. A. Thompson. *Parallax* 11(1): 6–9.

Gilbert-Walsh, J. 2000. Broken imperatives: The ethical dimension of Nancy's thought. *Philosophy and Social Criticism* 26(2): 29–50.

Heidegger, M. 1993a. Letter on humanism. Trans. F.A. Capuzzi and J.G. Gray. In *Basic writings*, ed. D.F. Krell, 214–265. London, Routledge.

———. 1993b. *Sein und Zeit.* Tübingen: Max Niemeyer.

Hutchens, B.C. 2005. *Jean-Luc Nancy and the future of philosophy.* Chesham: Acumen.

Kant, I. 1993. *Grounding for the metaphysics of morals.* 3rd ed. Trans. J.W. Ellington. Indianapolis: Hackett.

Lacoue-Labarthe, P., and J.-L. Nancy. 1997. *Retreating the political,* ed. S. Sparks, trans. unspecified. London/New York: Routledge.

Morin, E. 2007. Restricted complexity, general complexity. Trans. C. Gershenson. In *Worldviews, science and us: Philosophy and complexity*, ed. C. Gershenson, D. Aerts, and B. Edmonds, 5–29. Singapore: World Scientific.

———. 2008. *On complexity*. Trans. S.M. Kelly. Cresskill: Hampton Press.

Nancy, J.L. 2003a. Originary ethics. Trans. D. Lange. In *A finite thinking*, ed. S. Sparks, 172–195. Stanford: Stanford University Press.

———. 2003b. Concealed thinking. Trans. J. Gilbert-Walsh. In *A finite thinking*, ed. S. Sparks, 31–47. Stanford: Stanford University Press.

———. 2003c. Shattered love. Trans. L. Garbus and S. Sawhney. In *A finite thinking*, ed. S. Sparks, 245–274. Stanford: Stanford University Press.

Preiser, R., and P. Cilliers. 2010. Unpacking the ethics of complexity: Concluding reflections. In *Complexity, difference and identity*, ed. P. Cilliers and R. Preiser, 265–287. London/New York: Springer.

Raffoul, F. 2008. Derrida and the ethics of the impossible. *Research in Phenomenology* 38(2): 270–290.

Steinberg, J. 2015. Why I'm moving back to South Africa. *Buzzfeed,* 18 February, 2015. http://www.buzzfeed.com/jonnysteinberg/why-im-moving-back-to-south-africa#.ljONWykJBv. Accessed 23 February 2015.

Wittgenstein, L. 1921. Logisch-Philosophische Abhandlung, ed. W. Ostwald, *Annalen der Naturphilosophie 14.*

———. 2001. *Tractatus logico-philosophicus.* Trans. D. F. Pears and B. McGuinness. London/New York: Routledge.

Van Tongeren, P. 2008. Four types of theories of ethics: An encyclopedic synopsis. In *Woorden & daden. Een inleiding in de ethiek*, ed. A. van Melle and P. van Zilfhout, 173–180. Amsterdam/Heerlen: Boom/OUNL.

Von Foerster, H. 1990. Ethics and second-order cybernetics. Opening address for the international conference, *Systems and family therapy: Ethics, epistemology, new methods.* Paris, 4 October 1990. http://web.stanford.edu/group/SHR/4-2/text/foerster.html. Accessed 16 March 2015.

Woermann, M. 2013. Moral imperatives for an immanent world: Investigating the pragmatist, deconstructive, and ante-philosophical positions. *South African Journal of Philosophy* 32(3): 275–284.

Woermann, M., and P. Cilliers. 2012. The ethics of complexity and the complexity of ethics. *South African Journal of Philosophy* 31(2): 447–463.

Wood, D. 1999. The experience of the ethical. In *Questioning ethics: Contemporary debates in philosophy*, ed. R. Kearney and M. Dooley, 105–119. London: Routledge.

Chapter 7
Philosophical Complexity Revisited

At the dawn of the twentieth century, it was already clear that, chemically speaking, you and I are not much different from cans of soup. And yet we can do many complex and even fun things we do not usually see cans of soup doing.

- Philip Nelson (2003), Biological Physics: Energy, Information, Life

Abstract
In this final chapter, the notion of philosophical complexity is revisited and further explored in terms of the prominent philosophical insights gleaned over the course of the study. It is argued that systemic openness is the defining feature that distinguishes complex systems from non-complex systems. As such, the concept of opening constitutes the theoretical frame through which the post-structural insights are interpreted and translated in this chapter. Opening gives rise to three paradoxes that define complex systems, namely that these systems are both ordered and disordered (and are constantly in a state of decay and renewal); complex systems are autonomous from, as well as dependent on, the environment; and, the identity of complex systems is both exclusive and inclusive. Each of these paradoxes is explored at the phenomenological level.

7.1 Introduction

A number of complex philosophical ideas were examined in this work. A good starting point for translating these ideas into the language of complexity is to identify the essential characteristics of complex systems in order to establish to what extent the post-structural insights support, and add to, our understanding of these characteristics. This can be achieved by answering the question: How do cans of soup differ from human beings? From the perspective of complexity, I would argue that the best answer to this question is that human beings are open systems, whereas soup cans are not. The reason for this answer is because the system 'has replaced the simple and substantial object, and it rebels against being reduced to its elements' (Morin 1992: 97; italicised

M. Woermann, *Bridging Complexity and Post-Structuralism*,
DOI 10.1007/978-3-319-39047-5_7

in the original). The actual chemical composition of objects is thus secondary to their systemic identity. Moreover in complex systems, this systemic identity is co-constituted by the environment. Humans are therefore open to environmental inputs, whereas soup cans are isolated systems and preserve our food by virtue of the fact that they seal off the environment. The notion of opening is thus critical to understanding complex human systems. Morin (205) argues that '[f]ar from dissolving existence, [the concept of opening] reveals it; far from enclosing existence, it opens onto it.'

Morin (98) notes that a commonplace definition of a system is 'an interrelation of elements constituting an entity or a global unit.' An essential characteristic of systems that is missing from this definition is that of organisation. The importance of systemic organisation was first highlighted by Saussure, whom Morin calls a systematist, rather than a structuralist. Saussure (1931 in Morin 1992: 99) defines a system as 'an organized totality, made up of interdependent elements holding together and not able to be defined except one by the other in function of their place in this totality.' As mentioned in Chap. 2,[1] organisation is the means by which the interrelated elements are linked to a totality. As such, Morin (99) offers an updated definition of a system as '*a global unity organized by interrelations between elements, actions, or individuals.*'

Organisation not only explains how interrelated elements give rise to systems, but also how systems relate to their environments. Morin notes that, as with many definitions of systems, the notion of organisation is often neglected in discussions pertaining to systemic openness. Standard descriptions of systemic openness concentrate on inputs and outputs but fail to link the entry point to the exit point, in much the same way as standard definitions of systems fail to link interrelated elements to the emergence of a *unitas multiplex.* In order to correct this oversight, Morin argues that we should 'consider the organizational character of opening' (197).

The importance of organisation was also highlighted in the post-structural positions that were investigated in this work, albeit that the term *organisation* was not used. In Derrida's philosophy, the workings of *différance* link interrelated components in a network of meaning (which Derrida refers to as the text). The aneconomic and the economic dimensions of *différance* together give rise to Derrida's *aporetic* logic, which relates meaning to non-meaning, and which thus constitutes an organisational opening in the system. In Nancy's work, the notion of praxis—i.e. beings making sense in communication with one another—is the organisational principle that defines our mode of being in the world (and this mode of being is one of exposition). Nancy further defines our ontological condition as characterised by finitude. The conduct of sense, defined in terms of the non-fixing of sense, thus constitutes the organisational opening in our immanent system.

In this chapter, Morin's work on systemic openness (and how it is linked to organisation in particular) will be discussed in more detail. He writes that 'the idea of opening is a very great and profound idea, which transcends the idea of system' (197). I believe that it is specifically in relation to this idea that post-structural insights can contribute substantially to the complexity discourse, thereby broadening and deepening our understanding of complex systems.

---

[1] See Sect. 2.5, feature 8.

7.2 Systemic Openness

In Chap. 2,[2] complex systems were defined as both open systems and operationally-closed and bounded systems. It was further argued that systemic opening can only be conceived of in relation to a corresponding closing. Morin argues that, when speaking of complex systems, the opposition is thus not between openness and closure, but between activity and non-activity. The thermodynamic opening is necessary so that the system can exchange matter and energy with the environment. However, in order for the system (which is endlessly degrading) to reorganise itself, closure is necessary. Complex, autopoietic systems are thus operationally-closed.[3] Morin (1992: 198) argues that 'opening and closing appear as aspects and moments of a reality simultaneously open and not open.' The interplay between openness and closure constitutes an organisational, rather than a thermodynamic, expression.[4] Hence, opening should also always be defined 'by its organizational character (and not exclusively by import/export)' (198).

Organisational opening is not only relevant to humans, but also to machines. Active organisation is necessary for production. Yet, the difference between humans and machines is that humans cannot stop being open whereas machines can. Morin explains that machines are only functionally open, which means that when a machine is at rest (or when it is non-active) it loses its virtue of opening, and consequently also its quality of machine. In other word, it becomes a mere object. Living beings, in contrast, are constitutively open-closed. This means that organisational opening should be considered not only at the productive level, but also at 'the generative level of the recursive loop, the production-of-self, integral and permanent reorganization' (206). Morin explains that unlike machines that are open in order that they may work, human beings are open in order that they may exist. Human beings are thus also ontologically and existentially open.

Organisationally-open systems are defined by three paradoxes. The first paradox is that such systems 'live from death, to die from life' (204). As previously noted,[5] organisation should be thought of in relation to a corresponding disorganisation. Entropy, disorganisation, and decay are thus as critical to life as order or negative entropy, reorganisation, and renewal. The second paradox concerns the fact that organisationally-open systems are both dependent and autonomous. This is because, for example unlike stars that feed off themselves, such systems depend on their environments. Morin (200) writes that 'the organization of internal interactions and the organization of external interactions… constitute the two faces of [what he terms] auto-eco-organization.' Auto-eco-organization means that the environment co-produces the system (dependence), yet the environment is also set apart from the system (autonomy). The third paradox stems from the second and concerns the identity of the system. If the environment is critical to, but different from, the identity of the system, then the system's identity is defined in terms of both inclusion and exclusion.

[2] See Sect. 2.5, feature 6.

[3] See Sect. 2.2.2.

[4] See Sect. 2.5, feature 7.

[5] See Sect. 2.5, features 7 and 8.

These three paradoxes will be examined in more detail in the following sections. Before doing so, it should be noted that the level of analysis employed will concern human beings, understood in the phenomenological—as opposed to the biological—sense. A reason for this is because, as expressed in the introduction to Chap. 1, the aim of the study is to develop knowledge of the human condition, understood from the vantage point of complex systems. A second reason is that the philosophical insights gleaned pertain primarily to the nature of our being in the world. The notion of (what Morin terms) *phenomenal* or *existential openness* constitutes a profitable framework through which to further interpret these insights.

7.3 Live from Death / Die from Life

It is a biological fact that our organism is constantly dying and rejuvenating. Morin (204) writes that '[e]verything that is open lives *under* the threat of death and *from* the threat of death.' One is therefore literally not the same person that one was a couple of years ago. The interplay between thermodynamic openness and organisational closure is what allows 'existence [to] feed on what eats its away' (204; italicised in original), namely energy and matter. Interesting however it is not only our living beings that depend on the generative loop between order and disorder or life and death, but also our phenomenological beings.

From a phenomenological perspective, it is the interplay of meaning and non-meaning that is the seedbed of life. As argued in Chap. 3,[6] Bataille was the first to recognise the importance of excess and waste; and, moreover, to recognise that entropy always accompanies meaning (i.e. the restricted economy generates the waste of the general economy). In Morin's (204) words: '[c]onsummation, as Bataille had admirably seen… expresses both the fullness of life and the activation of death.' Yet, although Bataille recognises that consummation gives rise to an excess that cannot be assimilated in an economy of meaning, he fails to productively think the relation between meaning and non-meaning (in his work, the relation amounts to a non-relation). In Derrida's work, the interplay of meaning and non-meaning, or (more accurately) the *organisational economy* that relates meaning and non-meaning, is profitably conceptualised.[7]

In focusing on the relation between the restricted and the general economy, Derrida demonstrates the manner in which the general economy resides in the restricted economy. The restricted economy is therefore both open and closed, as the environment is drawn into the system without being appropriated by the system. Moreover, the play of the general economy within the restricted economy introduces disorganisation and entropy as necessary for the production of meaning. It is again worth repeating Derrida's (1982: 26–27) own words in this context:

[6] See Sects. 3.2 and 3.6.

[7] See Sects. 3.4.1 and 3.4.3.

> This unnameable is not an ineffable Being which no name could approach: God, for example. This unnameable is the play which makes possible nominal effects, the relatively unitary and atomic structures that are called names, the chains of substitutions of names in which, for example, the nominal effect *différance* is itself enmeshed, carried off, reinscribed, just as a false entry or a false exit is still part of the game, a function of the system.

Différance is the organisational principle at play in Derrida's philosophy. To recall[8]: *différance* is characterised by both an economic and an aneconomic dimension. The economic dimension concerns a regulated economy of difference and deferral—an economy, in other words, where order and organisation have gained the upper hand on disorganisation and entropy. The aneconomic dimension, on the other hand, refers to 'an expenditure without reserve... that apparently disrupts every economy' (19). The aneconomic dimension introduces entropy, disorder, and disorganisation into the system, which both enables 'nominal effects [and] the relatively unitary and atomistic terms that are called names', and renders the internal integrity of each concept and name as *a priori* ruined. These two dimensions serve to define Derrida's *aporetic* logic, which was illustrated at the hand of a number of examples in this work including the present and the gift, law and justice, ethics and politics, conditional and unconditional hospitality, animal/being, and the foreigner. Although these examples will not be revisited at this juncture, it is again worth noting that the two opposite and antagonistic realms of the economic and the aneconomic require us to simultaneously engage in the codification and the deconstruction of meaning. In other words, *différance* demands that we engage in a complex thinking.

In light of both Bataille's and Derrida's insights, the dictum 'to live from death, to die from life', can be re-inscribed in the phenomenological realm as 'the creation of meaning from non-meaning, non-meaning from meaning'. The first part of the formulation puts one in mind of the enabling force of *différance*. In Morin's (1992: 204) terms, it concerns 'death... from the inside (disorder in the reorganizational process).' The latter part of the formulation draws attention to both Bataille's insight that consumption produces excess and waste, and Derrida's injunction to deconstruct in order not to pass over the waste, but to try to wordlessly account for the inassimilable outside within the order of meaning. Deconstruction thus generates an awareness of the general economy.

7.4 Autonomy/Dependence

7.4.1 Forms of dependency

From the vantage point of complexity thinking, the economy of meaning does not depend on a metaphysical outside (Plato's Forms, for example). The outside merely refers to that which cannot be accounted for in our systems of meaning. One can however speak of our ecological dependencies in relation to a very real outside,

[8] See Sect. 3.4.1.

namely that of the world that we share with other people. As has been stressed numerous times, our physical beings require environmental inputs in the form of energy and matter in order to exist. Similarly, from a phenomenological perspective, we draw from one another and from the environment, even though our identities are also set apart from the environment. Interestingly, Morin (1992) uses the image of a mouth when speaking about the paradox of autonomy/dependence. He (204) writes:

> Ecological/existential opening is simultaneously the *mouth* by which life feeds its own existence and the haemorrhaging *breach* of its dependence and incompletion. The mouth is a breach, and the breach is a mouth.

This citation puts one in mind of Derrida's work on eating, which—to recall[9]—draws attention to our dependencies on other beings, who are themselves part of a subject's environment. Morin (201) reminds us of the richness and the importance of our exchanges with the environment in general in writing that:

> the environment... is much more than a reserve of food, still more than a source of negenentropy wherein being draws organization, complexity, information; it is one of the dimensions of life, as fundamental as individuality, society, the reproductive cycles.

In light of this expanded view of environmental dependencies, it is also useful to recall that Derrida's view of eating includes both the real and metaphorical assimilation of the other. In other word, it concerns all the ways in which we assimilate the singular other in order to live. This includes assimilation through eating, suckling, and sex (which are all literal forms of exchanging matter with the environment); as well as through words, sights, sounds, and touch (which denote more abstract forms of information assimilation). The metonymy of eating thus serves as a good example of our exchanges with the environment in general, and with other beings in particular.

7.4.2 Non-closure as a Condition for Generating Meaning

The metonymy of eating also introduces the economy of need and desire, which marks the closure of metaphysics, and which is premised on our need for the other. Through virtue of this economy, we become aware of the non-closure of life: autonomous life requires dependencies; it must continuously be supplemented by inputs from the environment (including from others). In Morin's (1992: 202) words: 'it is in... dependence that the autonomy of... beings is woven and constituted.' Krell (1988: 9) offers the following freestyle description of this bold economy, which highlights both our need and desires for others and the ways in which we depend on one another:

> engorgement of the breast with milk or the alimentary canal with food or the penis clitoris and lips with blood as experiences of the voice's site and situation the possibility of a projective and proprioceptive philosophy taking the measure of its line of thought from the breath that enlivens it and the flesh that informs it as a celebration of orality devouring the

---

[9] See Sect. 5.3.

> (m)ilk of the (m)other in full voice or rapt in silence as tumescence of the vocal chords here too what hegel called the vibrant blood but also the patient detumenscence of the stylo and its (b)analities no longer spirit's (s)tool but an opening onto the wor(l)d engorgement as the general economy of extravagance producing supplements of presence and disseminating signs of absence in defiant affirmative deferral of death

From a phenomenological perspective, the fulfilment of our desire for a fully present meaning is always deferred. Yet, this is exactly what drives meaning (and, hence, life) forward. This is because equilibrium (in this case, fully present meaning) results in system's death.[10] The economy of need and desire functions in a state of disequilibrium; or—in the words of Krell—it functions 'in defiant affirmative deferral of death'. Our dependencies on the environment and on others are thus the source of organisation and autonomy, as well as of disorganisation and *différance*. Together, these counteracting movements support the life force. Morin (1992: 204) states that '[a]ll wealth, consequently, is founded on insufficiency, all satisfaction on deficiency, all presence on absence, all present on the imperfect, I mean the non-perfect.' He further argues that recognising this insufficiency constitutes 'the origin of need, uneasiness, search, desire (which is not a reality first risen from who knows where, but a consequence of opening), [and] love' (205).

The above argument can be further elaborated at the hand of Nancy's ontology. As argued in Chap. 3,[11] Nancy's understanding of being is constituted by organised relationality. In his terms, existence is community in communication or beings-in-relation, and nothing more. Consequently, existence is constrained and restricted by itself; there is nothing beyond existence. As such, Nancy supports an open immanence. This view affirms Morin's assertion that, when speaking of complex systems, the opposition should not be thought of in terms of openness and closure, but in terms of activity and non-activity.

The activity of the system, i.e. being's organisation through virtue of communication, is a dynamic and ongoing process. In Nancy's (2003a) terms, it denotes finitude or the non-fixing of signification as the proper mode of being's conduct of sense, which is also nothing other than concealed thinking. Nancy argues that concealed thinking[12]—which in Morin's language constitutes the need, uneasiness, search, desire, and love that is borne from opening—thus necessarily has 'something of the uncompleteable about it' (34). Concealed thinking is related to some form of alterity; or, as stated by Morin (1992: 204), '[t]he truths of the existent are always incomplete, mutilated, uncertain, since they depend on what is beyond its frontiers.' Nancy (2003a) identifies the important challenge in this regard as 'one of knowing (or of not-knowing…) how to think a thinking that is still a thinking even when its content is not-knowing' (36). Thinking is therefore necessarily transgressive, incomplete, excessive, and open. And, in much the same vein as Morin argues that love is a consequence of organisational opening, Nancy states that concealed thinking is love.

---

[10] See Sect. 2.4.

[11] See Sects. 3.5 and 3.6.

[12] See Sects. 3.5.2 and 6.3.

As previously mentioned in a footnote,[13] Nancy (2003b) writes that the formulation 'thinking is love... can... begin the quest for an ignored essence of thinking for which we lack any evident access' (247). This is because framing thinking in terms of love draws attention to both our preoccupation with thinking (in philosophy, for example) and the elusive character of thinking. To elaborate further: Nancy argues that despite the fact that philosophy inscribes 'thinking is love' as 'the head of its program' (249), philosophy can never arrive at this thinking. The reason for this is that philosophy remains trapped within a dialectic, and is therefore unable to think the heart of love which is unable to enter into the dialectic: 'it cannot be posed, disposed, and sublated in a superior moment' (254). In other words, although philosophy is concerned with the essence of love—the love of thinking—this essence is necessarily missed in philosophical thought, due to the fact that thinking as love 'misses by essence its own essence' (254). Yet, Nancy agues, it is precisely because philosophy always misses love (and therefore itself), that it can give itself again to thinking. In Nancy's words: 'in thinking, it calls forth once again this love that it is... like a missed rendezvous, calls again for its repetition' (255).

This description resonates strongly with Morin's (1992: 205) description of the yonder as the origin of love. It is the non-place where the incompleteness of meaning (or Nancy's concealed thinking) 'is what is going to blossom, worsen, ferment, become exasperated in human subjectivity.' Moreover, it is through virtue of reflecting on the yonder—or, more accurately and less intentionally, it is through means of the yonder posed to thought—that 'the mystery of existence will emerge fully in one of the ultimate tendencies of philosophy, under the accurate name of existentialism' (205).

7.4.3 Praxis as Means for Co-constituting Being and the Environment

The above analysis illustrates both our material and ideological dependencies. We have a measure of autonomy, but we are by no means self-sufficient. Recognising our dependencies and incompleteness, is—paradoxically—what gives rise to the inexhaustible process of making sense, which is also nothing other than existence itself. We are thus constituted through virtue of our dependencies. However, it is not only being that is co-constituted in relation to the environment and to others; the environment itself is co-constituted by beings. In this regard, Morin (1992: 202) writes:

> Every open being acts and/or reacts on its environment. Every productive activity has multiple, diverse, complex effects on the environment. Praxis transforms: exports are not the restitution of import, the fashioned is not the given. The outside is transformed under the effect of the actions, reactions, products, and by-products.

[13] See footnote 20 in Chap. 3.

This description should be read in both the literal sense (i.e. the manner in which we transform the environment through means of our activities), but also in a more figurative sense, as the preservation and the transformation of ethos. Nancy explores this latter understanding of praxis at the hand of Heidegger. To recall[14]: Heidegger (1993) relates the question of humanism (what man is) to man's action, conduct, or praxis. Praxis is an end in itself, since what is at stake is being as such: it is only through conduct that being forms 'an essential and "active" relation with the proper fact of being' (Nancy 2003c: 175). Nancy defines this proper fact of being in terms of a relation to sense. Knowledge of sense is equivalent to the action of sense, in that '[t]o be is to make sense' (175). This view of ontology holds the implication that the 'the human is no longer the signified of sense... but its signifier' (183). We thus form our home in the world through virtue of making-sense. Ethos, which should be understood as abode, constitutes the activity of residing. The preservation and transformation of ethos is the outcome of praxis and our mode of praxis is being-engaged in thinking.

Morin (1992: 203) notes that 'opening is an essential trait of every praxic organization.' Applied to the above, this statement highlights the openness of thinking that defines praxis, which is also nothing other than a concealed thinking. Indeed, Nancy (2003a: 47) states as much in writing that '[t]he concealment of thinking is its praxis: thinking that undoes its objects in order to become the thinking that it is: *we*, with one another and with the world (47). Part of the value of this citation lies in the link that it makes between the phenomenal world and the real world. Our mode of being in the world—which is defined by both dependency and autonomy—gives rise to a thinking that cannot be appropriated by knowledge, and hence to the knowledge that we cannot exhaust meaning. This logic is beautifully captured by the poet, Wisława Szymborska (2006: 79–80), in a poem titled 'A Note':

Life is the only way
to get covered in leaves,
catch your breath on the sand,
rise on wings;

to be a dog,
or stroke its warm fur;

to tell pain
from everything it's not;

to squeeze inside events,
dawdle in views,
to seek the least of all possible mistakes.

An extraordinary chance
to remember for a moment
a conversation held
with the lamp switched off;

[14] See Sect. 6.3.

and if only once
to stumble upon a stone,
end up soaked in one downpour or another,

mislay your keys in the grass;
and to follow a spark on the wind with your eyes;
and to keep on not knowing
something important.

7.5 Inclusion/Exclusion

The above analysis of the paradox of autonomy/dependence highlights the complex manner in which systems are organisationally related to their environments. Both the system and its environment are dependent on, but also distinct from, each other. Because of the dependencies produced by virtue of active organisation, the system cannot simply be severed from the environment without destroying—at least in part—the identity of both. Morin (1992: 202–203) summarises as follows:

> A phenomenal retroactive loop… unite[s] the living being to its eco-system, the one producing the other, and conversely… Which leads to a fundamental problem concerning the identity and the intelligibility of everything which has ecological opening.

Morin (201) argues that eco-dependent beings have a dual identity: 'an identity which sets them apart, [and] an identity of ecological belonging which attaches them to the environment.' This statement can be interpreted in terms of Saussure's and Derrida's differential understanding of meaning,[15] in which the distinct identity of a given signifier is contingent on the network of signifiers to which it belongs (identities are thus construed in relation to one another). This statement can also be interpreted in light of Nancy's singular-plural view of being,[16] whereby identity is defined as relationality in that each singularity is constituted by a plurality, and each self is already also an other.

Nancy's view of being leads to a particularly fruitful interpretation of Morin's (201) understanding of the frontier, which he describes as follows:

> Whereas we tend to consider frontiers essentially as lines of exclusion, the word frontier, here reveals the unity of the double identity, which is both distinction and belonging. The frontier is both opening and closing. It is at the frontier that the distinction and the linking with the environment is effected.

The paradox of inclusion and exclusion plays out on the frontier, as the manifestation of existence itself. Being is always already a being-*with*, and the frontier is the place of our compearance or our mutual exposition. Furthermore, Nancy (2003d) argues that it is in our compearance—which is also a communication—that 'the

---

[15] See Sect. 5.3.1.

[16] See Sect. 5.4.

sense of being as finitude of sense' (74) is enacted. Morin's (1992: 201) description of the frontier resonates with Nancy's ontology, in that he defines the frontier as 'a place of communication and exchange… [and] the place of dissociation and association, of separation and articulation .' The frontier thus marks the horizon of our experiences. It is the limit of existence, where existence is constrained or restricted by itself. Nancy (2003d: 76) writes that '[o]n the horizon something is constantly rising and setting', and this something is being or existence itself. Morin (1992: 204) moreover defines being as 'a transitive uncertain *be-ing* [Fr. *étant*] which always needs to re-exist and which disappears as soon as it ceases to be fed, maintained, reorganized, reoraganizing…'.

The argument that systemic openness should be considered in terms of active organisation finds its strongest expression in the discussion of the frontier of being. Being does not reside in essences, but is constituted via the active organisation that links self, others, and the world. Active organisation is also nothing other than an expression of an open horizon or immanence, whereby 'a being-there (*dasein*), a phenomenal "be-ing" [Fr. *étant*] [is defined as] an existent whose existence supposes (and opposes) its own beyond' (205).

7.6 Open Solidarities

It has hopefully been demonstrated beyond dispute that operational-closure does not negate the fact that '[u]s, you, me, we are radically open' (Morin 1992: 207). Yet, despite being's openness, we see that the type of knowledge that is most commonly in operation in the world today is closed knowledge. Morin (205) writes that:

> closed knowledge has everywhere destroyed or hidden the solidarities, the articulations, the ecology of beings and of acts, existence! Thus we have become blind to openings, so true is it that the most difficult to perceive is the evidence that a dominant paradigm hides.

In Chap. 4,[17] it was argued at the hand of Derrida's and Foucault's insights that knowledge is always established within a conceptual hierarchy, which is itself influenced by an overarching epistemic frame. The act of knowledge production is therefore, by definition, exclusionary and violent. This sheds some light on why disciplines have developed in isolation from one another, and according to the principle of disjunction. It also goes some way towards explaining why we often treat others with suspicion and mistrust, rather than in terms of a mutual solidarity. Yet, the status quo can change, as is illustrated by the following anecdote relayed by von Foerster (1990):

> I have a friend who grew up in Marrakech. The house of his family stood on the street that divide the Jewish and Arabic quarter. As a boy he played with all the others, listened to what they thought and said, and learned of their

[17] See Sects. 4.2 and 4.3.

> fundamentally different views. When I asked him once, "Who was right" he said, "They are both right."
> "But this cannot be," I argued from an Aristotelian platform, "Only one of them can have the truth!"
> "The problem is not truth," he answered, "The problem is trust."

Trust requires openness. It requires that we find ourselves in each other. This can only be achieved 'by considering ourselves, human beings, as an open race marked by the existential wide-openness in our beings, feelings, love, phantasms, ideas' (Morin 1992: 207). One way in which to stimulate this openness is by recognising that our own identities are impure: we are both open and closed, dependent on the environment and autonomous, inclusively defined and exclusively defined, singular and plural. The intruder always already resides within the walls of the city and within the heart of being. As such, it is imperative that we step away from closed knowledge. We 'cannot', as Morin (207) asserts, 'isolate or exclude the situation… the explanation must be dialogical and dialectical, linking the inside and the outside processes in a complementary, concurrent, and antagonistic way' (207). The explanation, in other words, must be the outcome of a complex thinking.

Engaging in complex thinking does not lead to the type of knowledge that can readily be accommodated within the closed boundaries of disciplines, models, and hierarchies. Rather, the antagonisms and over-determinations give rise to excess, and it is this excess (or non-knowledge) that constitutes the praxis that is thinking and existence itself. Nancy (2003c: 175) writes that '[t]hinking (and/or poetry) is… what, in all action, brings into play the sense (of being) without which there would be no action'. Poetry can thus act as a corrective to closed knowledge, precisely because, as the poet Czesław Miłosz (2003: 240–241) writes:

> The purpose of poetry is to remind us
> how difficult it is to remain just one person,
> for our house is open, there are no keys in the doors,
> and invisible guests come in and out at will.

Contending with complexity is not an easy feat. Otherwise put, it is also difficult *not* to remain just one person. Yet, this engagement is unavoidable for the simple fact that things are not simple. However, it is good news that the world is complex. It means that there is space in the system and that we have a chance. As with any opportunity, the opportunities that stem from complexity also carry risks. Of late, we have unfortunately seen too many of these risks become a reality: the closed manner in which knowledge, states, religions, races, and ethnicities are defined all too often leads to violence, terrorism, intolerance, racism, and discrimination. Yet, the wonderful thing about complex systems is that their courses are not linearly set. The past influences the future, but the future need not resemble the past. It is not too late to establish an open solidarity, in which we live with open hearts and open heads. It is not too late to form a praxis of ethical living, defined by wide-openness.

References

Derrida, J. 1982. Différance. In *Margins of Philosophy*. Trans. A. Bass, 1–28. Chicago: University of Chicago Press.

De Saussure, F. 1931. *Cours de linguistique générale*. Gèneve: Payot.

Heidegger, M. 1993. *Letter on humanism*. Trans. F.A. Capuzzi and J.G. Gray. In *Basic writings*, ed. D.F. Krell, 214–265. London: Routledge.

Krell, D.F. 1988. Engorged philosophy: A note on Freud, Derrida and Différance. In *Derrida and Différance*, ed. D. Wood and R. Bernasconi, 7-11. Evanston: Northwestern University Press.

Milosz, C. 2003. Arts Poetica?. In *New and collected poems: 1931–2001*. Trans. R. Hass, 240–241. New York: HarperCollins.

Morin, E. 1992. *Method: Towards a study of humankind*, vol. 1, *The nature of nature*. Trans. and intro. J.L.R. Bélanger. New York: Peter Lang.

Nancy, J.-L. 2003a. Concealed thinking. Trans. J. Gilbert-Walsh. In *A finite thinking*, ed. S. Sparks, 31–47. Stanford: Stanford University Press.

———. 2003b. Shattered love. Trans. L. Garbus and S. Sawhney. In *A finite thinking*, ed. S. Sparks, 245–274. Stanford: Stanford University Press.

———. 2003c. Originary ethics. Trans. D. Lange. In *A finite thinking*, ed. S. Sparks, 172–195. Stanford: Stanford University Press.

———. 2003d. The unsacrificable. In *A finite thinking*, ed. S. Sparks, trans. R. Stamp and S. Sparks, 51–77. Stanford: Stanford University Press.

Szymborska, W. 2006. A note. In *Monologue of a dog: new poems*. Trans. C. Cavanagh and S. Baranczak, 79–80. Orlando, FL: Harcourt.

Von Foerster, H. 1990. Ethics and second-order cybernetics. Opening address for the international conference. *Systems and family therapy: Ethics, epistemology, new methods*. Paris, 4 October 1990. http://web.stanford.edu/group/SHR/4-2/text/foerster.html. Accessed 16 March 2015.

Author Index

M. Woermann, *Bridging Complexity and Post-Structuralism*,
DOI 10.1007/978-3-319-39047-5

Subject Index

M. Woermann, *Bridging Complexity and Post-Structuralism*,
DOI 10.1007/978-3-319-39047-5

CPSIA information can be obtained
at www.ICGtesting.com
Printed in the USA
LVOW02*1728011116
511202LV00001B/46/P